NITON PUBLISHING

BMW MOTORCYCLES

**The Postwar Range
with 1, 2, 3 or 4 Cylinders**

Roy Bacon

Published by Niton Publishing
PO Box 3, Ventnor, Isle of Wight, PO38 2AS

© Copyright Roy Bacon 1996

First published in 1982 in Great Britain by
Osprey Publishing Limited, 27A Floral Street,
London WC2E 9DP
Member company of the George Philip Group

A CIP catalogue record for this book is available from the British Library

ISBN: 1 85579 027 0

Original edition:
Editor Tim Parker
Design Roger Daniel

Reprinted by Bookbuilders Ltd., Hong Kong

Contents

Foreword by Reg Pridmore   7

Acknowledgements   8

1   Foundation laying   10

2   Reconstruction and revival   18

3   To the brink   30

4   New breed   48

5   Modern Times   68

6   Bright future   118

7   Solo racer   142

8   Sidecar racer   158

9   Speed records   168

10   Six Days and sidecars   178

11   *Wehrmacht und Polizei*   188

12   Specials and replicas   200

Appendices

1   Specifications   206

2   Colours   221

3   Model recognition points   222

Reg Pridmore aboard his winning R90S Superbike. Keen experts will notice the specification changes, amongst which are the deep sump, fork brace, caliper re-mounting, oil cooler vent and the rocker cover bashing of the twin plug head

Foreword by Reg Pridmore

To be able to cover such a vast number of changes and variations on BMWs over a period of some 35 years I should like to tip my hat to all concerned. This is a fine publication.

Discovering BMW for myself in the early seventies I find it hard not to reflect upon them as being definitely one of the best machines offered in today's marketplace. I should like to think that my five to six years of racing for the brand was instrumental in many of the refinements we are now seeing in today's production models. BMW endeavours to produce a machine to suit a wide range of riders whilst at the same time offering light weight, good handling and performance with a certain style and this surely has come about in the last decade.

Being a part of recent BMW racing history and working with people like Dr Adams (owner of Butler & Smith Inc.), also Helmut Kern (responsible solely for the race preparation of my B & S machine) has a certain significance for me. A Londoner racing a German marque in the USA at times created some very unusual and interesting circumstances. I found that the close association between myself and the factory representatives actually drew me even closer to the brand and the engineering and quality of materials made an even bigger BMW fan out of me.

Having raced and won at almost all the major tracks around the US plus an outing at Bol d'Or in 1975 on the Berlin twin, then a visit to the factory in the same year, I feel gratified with my BMWs.

My first big AMA national win came at Ontario Motor Speedway (California) in 1974 on the R90S creaming the might of factory Kawasaki rider Yvon Du Hamel and superstar back-up Steve McLaughlin on the big Z1s. I remember at Daytona outbraking one of my favourite heroes Dave Aldana (on the factory Norton at the time). Then those cool calculated wins at Laguna Seca and Riverside against ex-factory rider Gary Fisher and Steve McLaughlin, both on equally as fast BMW 1000s, with much Japanese competition being totally detuned along the way.

Seeing the photo finish at Daytona in 1976 must have been almost as exciting as the actual ride itself judging by the way I was received by the German contingent of spectators. After a mediocre start I worked my way up to first place by lap six, then finally down to the wire losing it by less than half a wheel (the pictures later revealed) to team mate McLaughlin alias 'Super-lips'. For me it was a race not easily forgotten. Nevertheless after the winner's circle I was lofted shoulder high by the Germans back to the garages losing my tee shirt to a good-looking German bird along the way.

Europeans are very enthusiastic. It's easy to get carried away talking about past road racing.

Of the many people that helped me take the BMW to its only National AMA Superbike Championship win in '76, I say thankyou.

A special thanks to Dr Adams, Helmut Kern, and Udo Gietle for making the BMW race programme possible.

Reg Pridmore
AMA Superbike Champion 1976,1977,1978
BMW R90S mounted in 1976
Ventura, California, USA
June 1982

Acknowledgements

I first rode a BMW in 1974 courtesy of Jeremy Frazer who was then the public relations manager of the UK importers. The machine, an R75/6, introduced me to the BMW world and to a man who set new standards in motorcycle promotion, and press motorcycling, for that BMW was the first such I borrowed.

It was to be followed by many others and thus I became aware of the pleasures and foibles of the Bavarian machine and its gearbox. In earlier years I had been impressed by sight of the black, flat twin on the road, with its quick and silent progress at a Thruxton 500. The solitary BMW pitted for petrol and rider change only, changed up at the same point on every lap and was as muted at the finish as the start. Every other machine was much noisier.

Writing this volume as the first in the Osprey Collector's Library series on a European machine posed different problems to earlier books on British bikes. The marque is still in production so a cut-off date has to occur while not too much of the earlier data was so readily accessible.

Fortunately I was able to obtain the help and assistance of a number of friends to whom I owe my thanks. As well as Jerry Frazer, now a PR consultant, my old racing adversary Mole Benn was able to provide some most useful material on the pre-war record attempts while Kay Thomson of Osprey Publishing helped by translating some German for me which cleared up a mystery on the wartime R75. In the course of writing the words reference was made to many sources and among the most useful were *BMW: A History* (Osprey), *BMW-Motorräder* (Ariel Verlag) and *Die BMW Kräder R12/R75 im Zweiten Weltkrieg* (Motorbuch Verlag).

The pictures came from a number of sources including the UK importers and I must especially thank Graham Pearce and Lynda Shoobridge of that concern for all their help and efforts to provide information and assistance. I am again indebted to the magazines for allowing me to borrow prints from their files and my thanks go to Bob Berry and Peter Law of *Motor Cycle Weekly* and Ian Beecham and Jim Lindsay of *Motorcycle Mechanics*. Other pictures came from the Institute of Motorcycling and Krauser.

As well as calling on old friends, I also made new ones while researching, these being Ron Slater of Hughenden M40 who sent me some pictures of his motocross outfit and arranged for one of the grass chair to be sent by Alan Whale. Ron also gave me some background information on both machines and the speed and efficiency with which he organsied the pictures reflects the reason for the success of his firm.

Nick Nicholls took the picture of Ron's motocross outfit and an early sidecar racing shot while one of the road model prints came from Don Morley (Allsport). If I have used any others taken by a freelance but unmarked, I can only apologise.

Finally I must thank English-born American racer Reg Pridmore for his foreword and Tim Parker, the editor, for his continued help in all matters publishing.

Roy Bacon
Niton, Isle of Wight
July 1982

Note This updated edition has been revised to include both the three- and four-cylinder models—long heralded and with a unique layout for production machines.

This title was further updated during 1996 to bring the BMW story up to date before reprinting it under my Niton Publishing imprint. This brought in the suspension advances and much revised engine of the 1990s together with changes to the appendices. As a result more pages had to be added to cope with the extra material, but I considered this a better move than trying to cram the original to make room. The page sidebars remain with the original book title 'BMW Twins & Singles' to provide a link from past to present. As always, I have to thank the staff at BMW for their help in providing material and photographs over the years for this task.

Roy Bacon
Niton, Isle of Wight
1996

1 | Foundation laying

When the new flat twin BMW was unveiled at the 1923 Paris Show it established a basic format that was to run on for well over half a century. Max Friz, the engineer who designed it, wrought well with his concept—so well that the BMW motorcycle came to mean the boxer engine and shaft drive while the blue and white quartered badge hardly needed the initials round the edge

The badge and the company both came from the German aircraft industry, for the roots of BMW lie not in the earth but in the sky. Thus the badge represents a whirling propellor shimmering against a blue sky and is based on a first design done in 1917 which ghosted in the engine and centre wing of a biplane behind the prop with the famous initials above. In 1920 came the first quartered badge, then as now with blue in the top left corner, but with yellow lettering and a scalloped edge. It was not until 1937 that the white letters appeared, and as late as 1963 when the present form was adopted with a change of letter type and subtle variations to the outer ring sizes.

The company came from a merger in 1917 but originated in 1913 when the Karl Rapp Motorenwerke München GmbH and the Gustav Otto Flugmotorenfabrik joined together to build engines principally for use in the air. In March 1916 the combine took the new name of Bayerische Flugzeugwerke AG, or BFW, and continued to build aircraft engines up to the end of the First World War. After that production of

Max Friz who joined the company in 1917 and set the high BMW standards

The Flink fitted with a Kurier two-stroke engine was the first company motorcycle

aircraft or their engines was forbidden and in 1919 all engine parts had to be destroyed under the orders of the disarmament commission. Thus six cylinder crankshafts went under the drop hammer.

Behind the technical scene considerable dealing had gone on to secure military contracts and through this two men came to Munich and the BFW works. One was a young Austrian flight officer Franz-Joseph Popp, who came to over-see the construction of a batch of engines, and the other was Max Friz who in 1916 was working with Daimler in Stuttgart. Friz was unhappy in his situation having been refused an increase in salary, and in the process of looking for a greener field contacted Karl Rapp with whom he had worked in the past. Rapp was uninterested but Popp, already a force behind the BFW scene, urged him to think again and so on 1 January 1917, Max took office in Munich. During that year Popp became more involved with the firm and in time became its technical director.

It was on July 20 in the same year, 1917, that Bayerische Motoren Werke GmbH, or BMW, was officially founded as the outcome of the various mergers over the years. During the year Friz designed a high altitude engine which became known as the BMW IIIa, and this proved to have an exceptional performance. As a result of its trials and the work of Popp behind the scenes on the business side BMW became an instant, undercapitalised success with an order book their facilities could never fulfil.

So on 13 August 1918, BMW GmbH, a limited liability company, became BMW AG, an Aktiengesellschaft or stock corporation with a capital of twelve million Marks. By October 1918 they had 3500 employees and were producing 150 engines per month, an immense change from the small-firm days of two years previous. BMW had become important.

The Armistice stopped all this at once and brought out the versatile leadership abilities of Popp and Friz who worked to keep the factory going on anything they could get to turn their hand to. They still turned their attention to the sky and, despite all injunctions, managed a little work on their high altitude engine while hiding plans and drawings in the heating ducts.

In 1919 flying by Germans was forbidden but in May BMW managed to get Franz Diemer into the air in a biplane fitted with the Friz engine. It reached a height of 9700 metres, a new world record and repeated this in June when 9760 metres was reached and the authorities confiscated all the related documents.

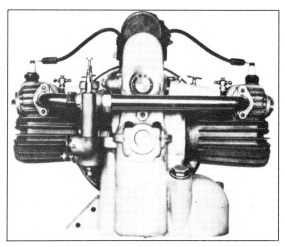

The M2B15 engine which was the first BMW flat twin and used by them and others

The Helios built by BMW using the M2B15 engine. Chain drive to rear wheel but crude brakes and forks

Popp and Friz had had their fun in this interlude but were soon deeply involved in the serious matter of finding work to keep the plant going. Much of this was for railway braking systems and it was during this time that the first contact was made with motorcycles in the search for diversification.

The first machine to be built by the company appeared in 1920 and was called the Flink. It used a 148 cc Kurier two-stroke engine which drove the rear wheel directly by belt. The cylinder was laid forward at an angle but its finning was arranged to be parallel to the ground while the magneto bolted to the back of the crankcase to form a vee shape. The engine was mounted in a loop frame with simple 'fore and aft' sprung forks and was started by pedalling gear connected to the rear wheel by chain. Footboards were provided well forward to complete a successful machine.

The Flink was followed by the first BMW flat twin engine, the M2B15, which was sold to companies such as Bison and SMW, and was used extensively by Victoria and by BMW themselves to produce a machine sold as the Helios. The engine was of 500 cc with square bore and stroke, side valves, one piece heads and barrels and produced 6·5 bhp at 2800 rpm. Its camshaft was sited directly over the crankshaft with the magneto above it and the connecting rods had split big ends. Lubricating oil was carried in a large sump, and in all cases the engine was installed with the cylinders fore and aft.

The Helios was fitted with a gearbox and chain drive to the rear wheel but retained a dummy belt rim for the rear brake, the crude forks of the Flink and no front brake. Some models had a speedometer whose drive box was clipped to the fork leg and driven by a band from the front hub.

It was not a very good motorcycle and BMW did not make many of them. Max Friz had high standards for his engineering and the Helios fell short of them. As a temporary expedient it could be tolerated, for a permanent design something different and much better was required.

And so the R32 was conceived by Max and turned into detailed parts by his assistants Martin Stolle and Franz Bieber. Stolle was not there to join in the excitement in Paris in 1923, for he had moved on to Victoria to develop their own flat twin engine, but Bieber was and went on to race BMW in 1924.

The first true BMW had the flat twin engine with cylinders mounted across the frame, unit construction of gearbox and shaft drive to the rear wheel. The essentials were not to change. The engine design was akin to that of the M2B15 with square dimensions and on a 5:1 compression ratio produced 8·5 bhp at 3300 rpm. The cylinders were still cast in one with their heads and retained the longitudinal fins despite their transverse installation. Valve caps were used to enable the valves to be fitted and the mixture was supplied to the inlets from a BMW carburettor.

The three-speed hand-change, gearbox was bolted to the rear of the engine and drove via a universal joint and exposed shaft to the rear bevel box. The frame was a duplex tubular construction with the right chainstays clamped into the rear bevel housing, while the left were formed with the rear spindle lug attached to the connecting bend. At the front were forks of the trailing link type controlled by a single leaf spring and small friction dampers. On the first model the speedometer was driven by band from the front hub, but there was no front brake and only a dummy belt rim at the rear. For the second series a 150 mm diameter drum brake was fitted to the front hub.

The R32 set the pattern without setting the world on fire with its performance for this only amounted to about 60 mph at a time when one of the better contemporary English machines was sold with a certificate of a 75 mph minimum. Even so it was soon running in German races and in 1924 Bieber collected the national title. This was the start of a run of success, for the 500 cc BMW held that title to 1932.

1925 saw the appearance of two new models, a 250 single and an ohv 500. The habit of building a single cylinder version of the BMW twin pattern was to last for over 40 years and all followed the same lines. From the start ohv operation was used and the valve gear was fully enclosed, in nearly every case, a decade or more before some other makes. The single cylinder was mounted on top of the crankcase as though this was a twin turned on end with one cylinder removed. The inlet, exhaust and ignition were tailored to suit the one lung but from the crankcase back the singles were a repeat of the twin design concept.

In practice on that first model, the R39, there were some unusual features. Not only was the valve gear enclosed but the valves worked in a separate light alloy head and the cylinder was cast in one with the crankcase and fitted with a pressed-in liner. A three-speed gearbox was used with its hand lever quadrant bolted to the head and barrel, while at the rear the bevel box was smaller, lighter and clamped to chainstays not parallel as on the R32 but pointed at headstock and under crankcase so they ran straight as did the front downtube.

The frame was still duplex, a type hard to avoid in the BMW concept and had trailing link front forks. The front brake was a drum but the other was an external contracting device which worked on the output shaft at the back of the gearbox.

The second new model was the R37, a machine designed for racing and fitted with a 500 cc ohv engine. This had the cylinder finning in line with the air flow and separate cylinder heads but otherwise was very similar to the R32 in engine, gearbox and cycle parts. As with all the early twins the kickstarter pedal was on the left and swung in the conventional fore and aft direction. It was not until 1928 that the transverse pedal, still on the left, appeared where it was to stay.

During the twenties the BMW continued on its motorcycle theme, while alongside it the aviation side had reappeared in 1923 and the company began to examine the motor car world. At first this resulted in the M2B15 engine being fitted to the Mauser three-wheeler, and another exercise in 1922 saw the same engine mounted in a Tatra. A little later prototypes were built using a watercooled flat twin engine but these too failed to come to anything. Then in 1928 BMW took over the Dixi works at Eisenach and had themselves a ready made small car to sell.

On the motorcycle front the 250 and 500 cc models were joined by a pair of 750s in 1928, one with side valves for touring and the other more sporting with ohv but still retaining the external brake on the gearbox. It was the latter model which joined its 500 cc brother on the race tracks in Germany and by then both carried superchargers mounted above the gearbox. BMW had used this avenue to more power since 1926 and it made their racing bikes some of the fastest in use.

Unfortunately the chassis side did not match the power output so that they were mainly successful on fast circuits but had to give best to the more nimble English and Italian machines on the slower and more twisty ones. In 1930 they pulled out of road racing for a while as record breaking had caught their attention, and Ernst Henne began collecting them for BMW in 1929. He pushed the absolute figure up nearly every year until 1937, when his speed became a mark not to be disturbed for 14 years.

In the early 1930s the road models took on a new appearance with the general adoption of pressed steel frames and substantial front mudguards often heavily figured with white lining on the black paint to create the effect of panelling. Technically the frame, forks and brakes remained little changed from the R32, although all were more substantial and up to the performance of the engines. The result was a machine whose appearance was cumbersome and heavy

Top **Start of the line—the 1923 R32 with all the BMW features of the next half century**

Bottom **An ugly BMW R17 of the mid 1930s; 750 cc and overhead valve and the 'square' frame of the R12**

The R71 with teles, plungers and side valves as built from 1938 to 1941. Nice matching sidecar

R16 of the very early 1930s. So called 750 ohv sports engine with Amal carburettors. BMW's curious front suspension

and which failed to compare well with other designs of the times. Teutonic in nature, it reflected back too much to the early twenties and the first BMW with a suggestion of an attempt to mask this with steel panels.

Within three years all this was to change and the first move came in 1935 when the two 750s, the side valve R12 and ohv R17, were fitted with telescopic front forks with hydraulic damping. While not the first motorcycle front suspension to depend on two pairs of sliding tubes, they were the first such fitted to a production machine that included hydraulic damping and were totally at one with those in use nearly half a century later. The R17 was also the most powerful and fastest road BMW to that date with 33 bhp at 5000 rpm from its very short stroke, 83 × 68 mm, engine. In the same year a factory racer came back onto the scene fitted with a new form of the same flat twin engine. This was of 500 cc with twin overhead camshafts to each head and supercharged by a blower fitted to the front end of the crankcase. The same blown engine was also used in the ISDT that year for it proved to be both fast and reliable.

While the change to telescopic front forks was a very notable step forward from the old trailing

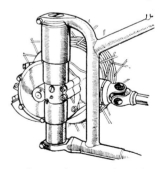

1939 rear plunger spring unit and the drive shaft universal joint

links, it had a surprisingly small effect on the appearance of the machines which still retained the pressed steel frame, hand gearchange and footboards of the past. It was 1936 when the real transformation took place and with the aid of the new forks BMW looks jumped forward to a standard that held good into the mid-1950s.

The machine that brought the transformation was the R5, a 500 cc ohv model with a new design of engine. This had two camshafts mounted above and to each side of the crankshaft to keep the pushrod length short and, it was wryly noted, these were chain driven. Hairpin valve springs went under new and larger rocker covers and on a compression ratio of 6·7:1 the engine produced 24 bhp at 5800 rpm. It drove a four-speed gearbox and this was controlled by a foot pedal on the left which pointed to the rear from a pivot below the barrel and was coupled with a rather long link to a short lever on the side of the box. There seemed little reason why a direct pedal could not have been used, as it was in due course.

There was not total confidence in this new mechanism for selecting the gears so a second lever was provided on the right side of the box, a short vertical stub, terminating in a ball for hand control, and this could be used in an emergency or by those who preferred the older way of doing things.

The revised engine and gearbox were welcome, but it was not they which brought about the transformation, it was the frame. Duplex and tubular it was welded not brazed and gave the R5 a sleek, light appearance that was no illusion for the machine was a good deal lighter than the R17. Somehow it looked altogether so much more modern with a line that was to run for two decades. Only at the juncture of the rear bevel box and its circular ring support pressing was there a glimpse of the past, and even this was not to last for long.

In 1937 the works racing bikes were fitted with rear suspension, while both single and twin cylinder road models appeared with the new type of frame which showed off the smooth lines of the engine and gearbox unit well. While the twin used the R5 frame, the single was built up with the main tubes being flattened and bolted to the headstock instead of being welded. The pressed steel frame was still retained for a cross country model and preferred for a military version as it gave such good all round protection to the power unit in the event of a minor spill. In the same year the air cleaner housing was incorporated into the top of the gearbox casting to start another long running BMW tradition.

1938 brought the final main change to the chassis with the introduction of plunger rear suspension on the R51 which replaced the R5. With it came three other twins of 600 and 750 cc, the last having the final side valve engine that BMW were to build. To accommodate the movement of the rear wheel the drive to it had to be modified and with its new mounting it completed the 1950s look of the machine. This was augmented by the adoption of a short direct gear pedal, a toolbox set in the top of the petrol tank, and a saddle with its suspension spring mounted between the frame tubes.

On the competition side, the blown twin won the European title in 1938 and the Senior TT in 1939, although the Gilera relieved them of the title that year. Then the continent was plunged into war and all production turned to things military.

At BMW these included the remarkable R75 motorcycle, a special sidecar machine built for all conditions from the desert to the Russian snows, although the other models continued to be produced up to 1941. Car and motorcycle production then stopped while the aircraft engine side expanded to meet the needs of the Luftwaffe, and went on to build a jet engine in 1943.

So the foundations were laid, but they were to be well rocked before another motorcycle was built with the blue and white badge.

2 | Reconstruction and revival

When the war ended in 1945 the BMW plants were in a poor state of health. The Munich works had been so badly damaged in air raids that it had closed. The Spandau plant in Berlin was damaged and at that time was in the hands of the Russians, who dismantled all the engineering facilities there. Eisenach was irretrievably lost, for it lay in the Russian sector. Only at Allach on the outskirts of Munich did BMW have a factory capable of doing some sort of work.

It had been engaged on the jet engine development and aircraft engine production, and was badly bomb damaged and then subject to looting until placed under American guard. It was the US military who in fact gave the Allach factory its first postwar work to repair their vehicles and so a small workshop was established. Then came reparation, and confiscation and demolition of the remaining plant seemed to be inevitable. Much of the factory machinery was crated up and sent on its way, but a small nucleus of BMW workers formed at the plant as old employees made their way there each day, and they managed to hang onto some of it.

By 1946 a little order was coming out of the chaos and there was no longer talk of razeing the plant. Instead, workers cleared away rubble, pulled down dangerous buildings, and improvised doors and windows in others. New products appeared using the one material they had plenty of—aluminium. At first they made cooking pots, then building fittings, and then a

Right **BMW** *bicycle* **built in the early postwar years and having an aluminium frame**

Below **Prototype R10 with 125 cc two-stroke engine but still a BMW with flat twin layout and shaft drive**

two-wheeler bicycle with an aluminium frame. This was of conventional form although the bottom bracket was larger than usual.

Immediately after the war Germany was prohibited from building motorcycles but the desperate need for transport relaxed this stricture. At first this only allowed a limited number of machines of not more than 60 cc but by the middle of 1946 there were no restrictions on that size and a total of 10,000 could be built in the 60 to 250 cc capacity range.

BMW were not slow to take advantage of this and the man who inspired them to take this course was Georg Meier, the 1939 TT winner. During the war he had directed plant security for the company and welcomed the chance to get the works back onto two wheels.

The idea of motorcycles once more led to the construction in secret of a miniature BMW, the R10, with a 125 cc flat twin engine operating on the two-stroke cycle. This may have been out of context, but otherwise the model showed its ancestry only too well with unit construction, shaft drive, telescopic front forks, and plunger rear suspension. The machine did not go past the prototype stage but the idea did, in East Germany at Zschopau in the former DKW factory where first the IFA and then the MZ were built.

In practical terms, a simple single was the easiest course back into manufacture and had the merit that it could readily be doubled up into a flat twin once more. So, engineering drawings for a new 250 were prepared, even if not much metal could be cut at first.

The first signs of a renewal of the marque came at the Paris show held in October 1947 where some BMWs were seen, but these were rebuilt models incorporating some French fittings and not available to buy. At the same show came a report that the Russians were making BMWs.

The new 250 made its debut at the Geneva show in March 1948, this being the first postwar international exhibition attended by BMW, and was numbered the R24. It had strong links with the pre-war R23 with a very similar rigid frame built from tubes bolted together and carrying telescopic front forks. Both wheels had drum brakes and 19 in. rims spoked to the hub, that at the rear connecting to the usual bevel box.

The engine followed the R23 with its equal 68 mm bore and stroke, but otherwise was new. It was based on a barrel crankcase which carried the crankshaft with its bobweights in substantial mains, one fitted in the rear case wall and the other to a carrier plate.

The cast iron barrel was bolted to the alloy crankcase and was fitted with a light alloy cylinder head and flat top piston to give a compression ratio of 6·75:1. The camshaft was sited above, and to the left, of the crankshaft from which it was chain driven to lift tappets which slid in pressed-in guides. The pushrods were enclosed in separate tubes running from

The R24 of 1949 with bolted frame construction and rigid rear end of pre-war form

crankcase to head and reached up to rockers supported by pillars rising from the cylinder to a level above the head. The rockers carried conventional adjustors at their inner ends and controlled the valves retained by coil springs. The entire assembly was enclosed with an air gap between inlet and exhaust sides.

The engine was lubricated on the wet sump system and had a submerged pump driven by skew gears and a long shaft from the cam. It was fitted below a filter tray in the crankcase under the crankshaft and access was via a sump plate bolted to the bottom of the crankcase casting.

At the front of the crankshaft went a Norris dynamo along with ignition contact points and advance mechanism. At the rear went the single plate car-type clutch and this drove a new four-speed gearbox via a spring-loaded cam type shock absorber on the input shaft. As before, the gearchange was by pedal on the left but with the supplementary short hand lever on the right. The kickstart pedal was also on the left and worked

transversely as before.

The remaining details were simple and straightforward and much as in 1939. The petrol tank had a large diameter filler cap, then seldom found, and the toolbox set in its top. A saddle and rear carrier were provided, the front and rear mudguards were extensive and the headlamp shell accommodated the speedometer driven from the output shaft of the gearbox. The exhaust pipe was carried on the left and connected to a low level silencer, while the Bing carburettor had its air cleaner attached. A centre stand was fitted and the machine finished in the traditional BMW black with white lining for tank and mudguards.

Production was not due to start until 1949 but began in December of the preceding year and by May, 800 men were building 50 machines a week. They were helped by the extensive range of new machine tools they had managed to obtain to replace those lost in reparation and were hardly displeased that the older and less

21

efficient equipment was being used by their competitors. In the foundry they had an ample supply of light alloy billets in the form of old aircraft cylinders and so could use the material extensively. At first they sand-cast the large items, but quickly these joined all the others and were die-cast which gave accuracy, lighter sections, and less machining.

The R24 was only built for just over one year and in May 1950 BMW had a revised version, the R25 with some small engine changes and a new tubular frame without the bolted construction but with plunger rear suspension. This removed the pre-war appearance of the bevel box and allowed the use of interchangeable wheels. These had received one benefit of earlier work and had straight spokes. To meet the needs of the times the 250 was also offered with a sidecar sold under the BMW name.

On the twin cylinder side an indication of the factory's intentions was seen at the March 1949 Geneva show where an engine and gearbox unit was exhibited, it being a development of the pre-war R51 fitted with the new type cylinder head from the 250 and a similar gearbox. In November that year the capacity restriction on German machines was lifted and BMW announced their R51/2.

As expected, it followed the pre-war machine closely in both engine and chassis but did sport a pair of downdraught carburettors on the new cylinder heads. These repeated the pillar rocker mounting of the 250 and also the coil valve springs which replaced the earlier hairpins. The gearbox was patterned on the 250 design and had the same internal changes to improve gear selection.

On the chassis side the frame was, as before, constructed of oval section tubes electrically welded at the joints. The section was chosen to give the desired strength in the required direction and the welding gave neat, light joints. Plunger rear suspension was used and the telescopic front forks had two-way hydraulic damping. The brakes were large and stylish, the front mudguard deeply valanced, and the petrol tank had the toolbox set in its top. The saddle was linked to a spring set between the frame tubes, and a rear carrier was provided.

The BMW twin was back and made its show debut at Geneva in March 1950. In August they made a bid for US sales by appearing at the Trade Fair in Chicago where the sober and distinguished lines of the twin made a good impression. By the end of the year one twin came off the end of the assembly line for every two singles, and three out of four machines went to the German home market to be snapped up at once.

Late in 1950 *The Motor Cycle* road tested a R51/2, although the model was not then available in the UK. It proved to have a subtle blend of high performance and gentlemanly behaviour for, on the one hand it was good for 88 mph—a good speed in those days— while on the other it was quiet, completely oil-tight and ticked over slowly and surely. It also exhibited many of the BMW characteristics that were to ring down the years. The clutch freed well but took up rather sharply, there was a degree of harshness in the power delivery at low speed in a high gear, this all smoothed out as the speed increased without any vibration intruding, the gear change had to be taken slowly, the suspension was firm and well controlled, the handling was good, stable and light thanks to the low centre of gravity, and the footrests were well back as dictated by the cylinders and carburettors. Not much of which was to change for a long time except in degree.

Less usual on that road test was a front brake that lacked power despite its 200 mm diameter. The slow action of the twistgrip and generally heavy controls were found to be tiring but the mudguards were very effective, as was the main headlamp beam, while the dip had the sharp cut-off normal with European lights of the period. In all, the BMW came across strongly as a machine of quality possessing a universal appeal.

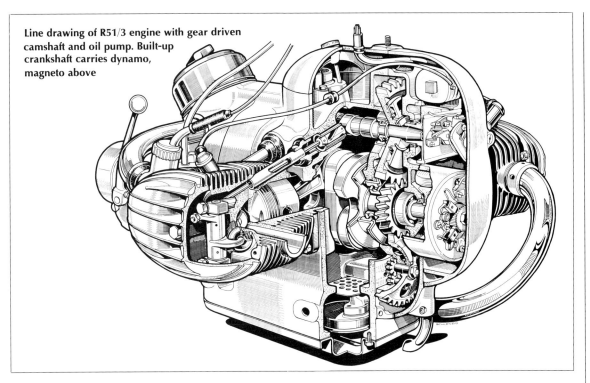

Line drawing of R51/3 engine with gear driven camshaft and oil pump. Built-up crankshaft carries dynamo, magneto above

The engine of the R51/2 seen at the Brussels Show in 1951. Chain camshaft drive, magneto on engine top

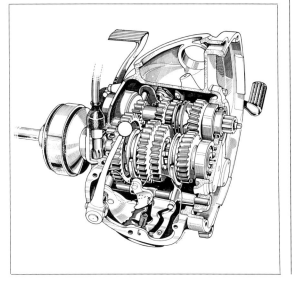

Cutaway drawing of gearbox fitted to R51/3 and R67/2 engines. 4 speeds, kickstarter, footchange and hand selector

For 1951 the 250 cc single continued as the R25/2 with minor changes, but more important was the appearance of the R51/3, with a much revised engine, and the first postwar 600, the R67. Both these twins used the same basic design with the smaller one keeping its 68 × 68 mm dimensions as before along with the same compression ratio, power output, and gearing. The new 600 had 72 × 73 mm bore and stroke and an actual capacity of 594 cc. Compression ratio was 5·6:1 and the power a little up on the 500 at 26 bhp at 5500 rpm. The gearbox and transmission were as the smaller model, but the rear axle ratio was raised for both solo or sidecar use.

It was in the layout of the crankcase that the main changes were made, for the camshaft chain went to be replaced by gears. At the same time the camshaft itself was moved back to a position immediately above the crankshaft from which it was driven by the single pair of helical gears. Below the crankshaft and also driven by gears went the gear-type oil pump set in the front main bearing housing, this being a substantial casting that spigoted to and closed off the end of the crankcase barrel. The pump was connected by a short pipe to a pick-up filter in the sump.

The front of the crankshaft carried the dynamo which was designed to be short and fat to suit the space, with the commutator and brushes at the front. Above it, on the end of the camshaft, went the magneto rotor, ignition cam and auto-advance with the stator bolted to the crankcase wall so that the ignition coil was at the top. The whole of the electrics, both dynamo and magneto, were enclosed by a single alloy casting completely smooth and free from holes. The electrical connections emerged further back with each plug lead exiting from the side through a grommet, while the low tension side came out at the top under the fuel tank.

The crankshaft was built up with roller big ends and ran in ball bearing mains. The rods carried conventional pistons, with small valve recesses in their crowns, which moved in iron

Above **The R68 sports model of 600 cc with combined saddle and pillion seat**

barrels bolted to the crankcase. The heads were light alloy and a pair of small tubes joined each to the crankcase and each contained a pushrod. The tappets worked in pressed-in guides and were extensively hollowed out for lightness to compensate for the additional weight of the longer pushrods. The heads and valves were as before but were enclosed by a single alloy cover which replaced the two which had been held in place by a strap. Each head was fitted with a Bing carburettor and these had curved inlet pipes to connect with the air filter set in the crankcase above the gearbox. The oil filler and dipstick was located on the left side of the crankcase just behind the cylinder, where it was to stay for at least three decades, out of the way and awkward to fill.

The transmission followed earlier BMWs with four-speed gearbox, foot pedal on the left and

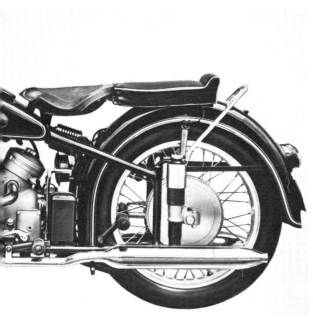

hand stick on the right, exposed shaft drive with flexible coupling and enclosed universal joint, and final drive unit with spiral bevel gears.

The chassis was that of the R51/2 with its telescopic front forks and plunger rear suspension. At that point the hubs were not full width and the rims were steel, but both the models featured all the usual BMW details and the distinguished black and white finish with the wheel rim centres in silver.

Motor Cycling had a 500 on test in May 1951 when it cost well over £400, or about twice the price of an English twin of the same size. Only a Manx Norton or Vincent Black Lightning cost more, so the UK demand for the German twin was small. For that price the owner received a

Below **A page from a BMW catalogue showing the 250 cc model also sold with a single seat sidecar**

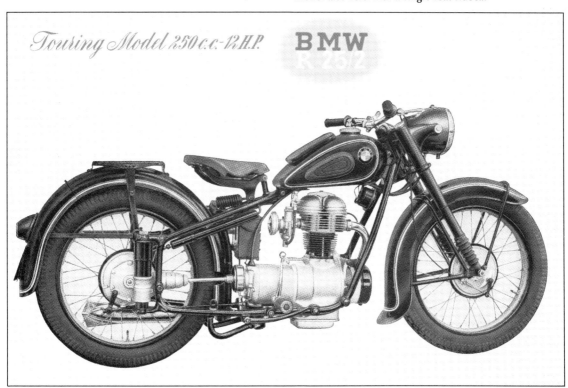

Touring Model 250 c.c-12 H.P. **BMW**

Right **A 250 with the rear wheel off showing its drive and brake shoes. Note slotted holes for mudguard stays**

Centre right **Engine unit of the R67/2 of 1954**

Far right **BMW saddle with single suspension spring mounted to frame. Tyre pump on frame tube of this twin**

Below **Engine unit of 1951 R51/3. Air cleaner differs from 600**

Below right **Drive shaft and rear wheel of the R51/3 model. Effective silencers**

machine that exemplified the BMW image of a luxury model of distinction and refinement for a gentleman of leisure who just might, on occasion, wish his quiet steed to cruise up to 80 mph. This it would do with perfect manners with a fuel consumption around 60 mpg and a constant tickover no matter how hard it had been driven.

There were minor criticisms as to the hardness of the suspension and the setting of some of the controls, while the gearchange was not to be rushed. The ignition could be turned off by a removable key which also worked the headlamp switch and was a common Bosch feature for many years, while a neutral light was provided in the headlamp shell alongside the charging warning light. Perhaps there were more criticisms listed than there should have been at the price but this was far more competitive in Europe and

especially in Germany. For those who could manage the price they found they had a very good motorcycle indeed.

In March 1952 a sports 600, the R68, made its debut at the Geneva show, always then a good barometer of the European motorcycle industry. The Swiss event was central to the main producers and was thus often used for launches for it was considered neutral ground and thus allowed prices to be compared on a really realistic basis.

The R68 used much from the existing twins but its compression ratio was up and the power output became 35 bhp at 7000 rpm. Externally the engine unit looked nearly the same with only a change to the air filter housing and cleaner rocker covers with simpler lines giving an indication. The chassis was that of the R67 but with a twin leading shoe front brake actuated in the simplest manner with the cable outer pushing one cam lever and the inner pulling the other. The mudguards were more sporting than usual on a BMW and the machine was fitted with

The sports R68 on show fitted with a high level exhaust system and twin leading shoe front brake

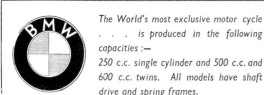

The World's most exclusive motor cycle . . . is produced in the following capacities :—

250 c.c. single cylinder and 500 c.c. and 600 c.c. twins. All models have shaft drive and spring frames.

Model 51/3

Now in stock at the following distributors:—

RAYMOND WAY of KILBURN Kilburn Bridge, London

KINGS of OXFORD   -   -   -   - New Road, Oxford

CLAUDE RYE - - - 895-921 Fulham Road, S.W.6

KINGS of BIRMINGHAM   - Bristol Street, Birmingham

JORDANS of HULL - - - - Story Street, Hull

KINGS of MANCHESTER 770 Chester Road, Manchester

Sole concessionaires for Great Britain:—
A.F.N. LTD., FALCON WORKS, LONDON ROAD, ISLEWORTH, MIDDLESEX

1951 advert for the marque that appeared in *Motor Cycling*. **Price not mentioned**

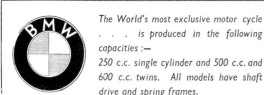

the linked saddle and pillion seat which was a partial step to the dualseat. The model was BMWs first 100 mph road bike.

The other machines in the range for 1952 had few changes. Both 500 and 600 twins received the new front brake and the larger R67 had its power raised a little to 28 bhp at 5600.

There were no major changes for 1953 but near the end of the year the 250 cc single was replaced by a revised model, the R25/3. This had a slightly improved power output aided by an air intake passage that curved up from the Bing carburettor to run forward in the tank tunnel to a filter, with strangler control, mounted under the front of the tank and immediately above the exhaust pipe. An unusual feature incorporated soon after the model's launch was to black anodize the alloy cylinder head. This was not done to enhance the appearance, but to improve the cooling which it did in a number of important areas.

The cycle parts had revised suspension with longer travel and hydraulic damping in the front forks. Both wheels were fitted with full-width light alloy hubs and straight spokes together with alloy rims of 18 in. diameter, smaller than before. The overall appearance was improved by these changes with the petrol tank lowered to shroud the rocker cover and the use of the saddle from the twins. The hand gearstick went but otherwise the model still had its German styling. It also still had the tools in the petrol tank compartment but in a revised position on the left side. The lockable toolbox lid was neatly hidden by the left kneegrip which was fitted to it and the assembly hinged down to reveal the large, quality kit. In its new position it was no less accessible to the rider but much less noticeable and tempting to the casual passerby.

The three twins for 1954 had their forks improved with two-way hydraulic damping and lost their upper spring covers which were replaced with rubber gaiters. They were also fitted with full-width light alloy hubs, straight spokes and alloy rims, except in the case of the R67. As this was built in the main for sidecar use it retained the steel ones.

Of the twins it alone continued on after 1954 for BMW were on the point of a major revision to their machines which had not changed very much in any major respect since 1936 and their first tubular frame. The 250 kept on into 1955 but the R67 became a sidecar machine for enthusiasts who preferred the plunger frame for that purpose. It continued in this form into 1956 but was then discontinued to bring the long line of plunger BMW machines to a halt.

3 | To the brink

As Germany entered the second half of the 1950s business looked great at BMW. The motorcycle side had reason to feel pleased and proud of their achievements for by late 1953 they had produced their 100,000th postwar machine and were running at a rate far in excess of pre-war days.

They also had a revised range of models to offer with new frame and suspension systems. By 1955 most large capacity motorcycles had turned away from plunger rear suspension to the swinging fork. Smaller machines were also following suit as the dictates of economy and tooling allowed. The beauty of the plunger system from the maker's viewpoint is the ease with which he can add the plunger housings to the end of a rigid frame with the absolute minimum of fuss or bother. He may have to increase the size of the wheel spindle to deal with the added loads, but the rest of the machine can continue as it was and still be built in rigid form.

Once the swinging fork is introduced many frame changes are needed and in turn these require numerous modifications to many other chassis parts. Nearly all of the major makers went along this route and used the exercise to update the rest of the machine, although this placed a heavy load on the various factory departments.

BMW managed to avoid this upheaval while incorporating a rear fork with ample movement and at the same time retaining one plunger-framed machine, the R67, for a year or two. They

The R69 typified the late 1950s BMW with Earles forks
and swinging fork rear suspension. This model is still
fitted with the combined seats

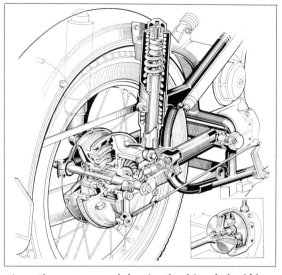

Above **The new rear end showing the drive shaft within the fork leg and the method used to attach the spring units to the frame**

Left **Front suspension with the Earles fork and friction steering damper**

did this by redesigning the frame in a manner that made few alterations. The top tube was still the only single one with all others duplex. The down tubes still ran down under the engine and back to the rear wheel but, instead of the abrupt corners and plunger housings, there was a smooth curve to bring each tube round and up to the rear end of the top tube. Into each loop went a slightly curved vertical tube to support the rear fork, and on top of each loop at the rear went support pressings for the rear spring units.

The frame was cross-braced above and below the fork pivot but elsewhere the engine crank-case did this duty. The rear fork itself was a rather special unit for BMW had decided to hide the drive shaft from the public eye by running it down the right fork leg. This allowed the bevel box with its crown wheel and pinion to be bolted directly to the end of the fork arm as an assembly. It also prevented the use of a normal

fork pivot so this job was done with two pivot pins which screwed into the frame. These allowed the fork to be centralised within the frame and also provided adjustment for the taper roller bearings it pivoted on. Movement of the fork was controlled by two spring units with hydraulic damping which fitted between it and the frame pressings. Each was adjustable for the load of a passenger by means of a short lever welded to the lower section.

At the front, BMW decided to break from tradition and use long leading links of the type called 'Earles forks'. At that time such forks with either long or short links were in use on small machines, but not often on larger. One or two road racing machines used them but in time most returned to the telescopic fork despite its inherent weaknesses.

The BMW fork was straightforward with the main tubes bent back from the lower crown to

On test with an R26 from 1956, a very pleasant if costly machine

engine units from the R51/3 and R68 respectively with small increases in compression ratio and power. Behind the flat twin engine went a new diaphragm clutch and behind that, a new three-shaft gearbox in an improved housing. It retained the four speeds and cam lobe shock absorber on the input shaft but was otherwise improved to speed up the slow gearchange.

The remainder of the cycle parts were much as before but among the changes was a deeper petrol tank with the toolbox hidden behind the left kneegrip as on the 250. A pivot saddle supported by a tension spring was still fitted and the mudguards, silencers, and general detail work and finish were of the usual BMW high standard. Wheels were of 18 in. diameter and the rims were in light alloy.

The Motor Cycle road tested an R50 in the middle of the year and found it quiet, docile and comfortable while capable of 90 mph, a good speed for a 500 cc twin at that time. While the new clutch freed well, the new gearbox still had a change that was difficult to make cleanly and quietly. The riding position was an asset to fast travel and the offset of the footrests to match that of the cylinders was noted as having no disagreeable effect. It was an expensive machine but a very good one.

During the year the twins were joined by a single cylinder 250 cc using the same form of front and rear suspension. This was the R26 with an increase in compression ratio and power together with the same gearbox ratios as the twin. It retained the earlier single plate dry clutch but in most other matters followed the twins. Curiously, the toolbox went back to the top of the tank while the air cleaner housing became a massive affair that also enclosed the battery.

For 1956 there was another new model, the R60, which was to replace the R67 discontinued late in the year. The R60 was thought of essentially as a sidecar machine and so used the R67 engine with its low compression ratio of 6·5:1 and reduced power output of 28 bhp at

run back to the pivot point. They were joined by a cross-tube just above this and the fork pivoted on adjustable taper roller bearings on a pivot pin. Two spring and damper units controlled its movement, while that of the fork assembly was held in check by a friction steering damper.

Both suspension forks carried new style full-width hubs with straight spokes and the twins used the twin leading shoe front brake controlled by the cable working on the two cam levers. The single retained the single leading shoe and its operation by the cable outer pushing the cam lever, while the inner end located in a lug on the brake backplate.

Initially there were two new models, the R50 of 500 cc, and the R69 of 600 cc. These used the

5600 rpm. It used the clutch, gearbox and cycle parts of the R50 and, although intended for sidecar use, was also available with a solo ratio rear axle. In that guise it made a very pleasant tourer.

The other machines continued unchanged for BMW never altered anything without good reason. A road test on the R69 carried out by *Motor Cycling* showed that there were few items in need of alteration anyway, although the gear change was still subject to the abilities of the rider. Top speed came out at 102 mph, while the front brake alone returned a braking figure few machines could match with both in use. Fuel consumption was around 70 mpg most of the time unless the model was cruised around the 90 mark with odd bursts up to 100.

And so the motorcycle range settled down, with the departure of the R67, to four models all using the same form of cycle parts. The 250 cc single, a 500 cc twin, and two 600s, one sporting and the other for sidecar use.

A very nice R60 with Steib sidecar, the latter not the usual and better known bullet shape, and in white

Behind the scenes the company was far from well and had been ailing for some time. As far back as 1953 there had been the first signs of a change in the needs and aspirations of the customers. The hectic days of the immediate postwar period were over and people were moving on from any form of transport at any price to be more particular. In each country a sequence was to occur of bicycle, moped, scooter, bubble-car, small car, bigger car, as times improved and prosperity returned. The timing of this sequence varied from country to country and astute manufacturers switched their output to suit. Motorcycle production ran alongside this sequence and as a form of transport was linked to it.

After the war the cry had been for production, and in Germany it was met with such determination that the economic miracle came about. But no-one called for any lessening of output and, by the middle 1950s, it was outstripping demand even though that demand was itself still increasing. With it came the inevitable move from two wheels to four and this caught BMW poorly positioned.

Above **The Isetta with twin rear wheels. Three adults could get in with a squeeze**

Below **Like this one**

"Just wait till I get out . . . !"

They built very good but expensive motorcycles and the cars were all too similar with large six and vee-eight engines far removed from most buyers' budget. Something else with more mass market appeal was needed, and in the search BMW explored many avenues. One result was the BMW scooter which used a 175 cc four-stroke engine and followed the pattern for such vehicles. It reached prototype stage about 1954 and was due to be released the next year, but was dropped. A smaller twin was the next rumour with the model based on the R50 and of 350 cc capacity, but this would still have been expensive and little more was heard of it. The firm had built a 250 cc twin as an experiment in 1939, but whatever the year the economics did not indicate this to be a solution.

Motorcycle sales fell dramatically. 1953 and

Geoff Duke plus passenger in the 600, a stretched Isetta with a second seat and side door too

1954 had seen about 27,000 units being sold each year but this figure slipped to 23,000 in 1955 and slumped to 15,500 the next year. 1957 was even worse with only 5400 units selling, and the workforce had to go onto part-time or be laid off. Motorcycling was about to enter a difficult period which ran through the 1960s, and all firms had problems during this time.

At BMW they sought to counteract the drop in motorcycle sales by moving into the bubble-car market. They did this by buying the rights to produce a small two-seater with front opening door which had been designed by Iso. This

became the Isetta and was first shown by BMW at Frankfurt in 1955. It had four wheels but the rear pair were so close together that they could count as one in some countries and so the vehicle could be classed as a three-wheeler with tax and insurance concessions. For other countries a version with a single rear wheel was built.

The Isetta was driven by an R25 engine fitted with a cooling fan and this located behind the

The 700 saloon which used the flat twin engine and helped to tide the company over the troubles of 1959

passenger on the right and drove the rear wheels via a four-speed gearbox. A tubular and box section chassis was used and most types were left-hand drive with the steering wheel attached to the entry door so it swung out of the way with it to facilitate entry and exit. While essentially a two-seater it could, and did, carry three adults on many occasions.

It was popular and sold very well but was unable to do very much for the firm's finances. It was later built with revised windows and also with a larger engine using the R60 dimensions, but still keeping the single cylinder. It remained in production until 1962 and was also built under licence in many other countries. It was cheap to buy and run, fitted into very tight parking spots and, like all of its kind, was doomed the day the Mini was announced.

In its wake came the BMW 600 which at first sight seemed like a stretched Isetta. In many ways it was for it retained the front opening door and rear engine, but between them went a

second seat with access to it by an additional door in the right side. It was a true four-wheel car with a rear suspension form to be used by BMW for many years.

The power unit was not the expected R60 but an R50 bored out to 74 mm and 585 cc. It was fed by a single carburettor and produced 19·5 bhp at 4000 rpm. It had fan cooling but the engine owed much to the motorcycle unit. The result was a small car that could reach just over 50 mph, and still BMW had not solved their financial problems.

The next step was another small car with a motorcycle engine, the 700. This at least looked like a car and was fitted with a bored-out 600 unit of 78 × 73 mm dimensions and 698 cc. In its original form it produced 30 bhp at 5000 rpm, but later ones went to 40 at 5700 while retaining the air cooling. It had a smart Michelotti designed body and was a half step towards filling the yawning gap between the motorcycles and Isettas at one end of the scale and the big 500 series cars that sold in such small numbers at the other end.

To fill that gap BMW needed a new middle class car and, without it, the two-wheelers, the bubbles, the promising 700 and the ailing aircraft engine side would all go down. By 1959 the situation was grim and the company was very close to failing. Investors came to Munich often in those days to see what was left to be picked up, but the spirit that had seen BMW through the dark days just after two world wars was still there and determined to keep the company afloat.

The 700 was the key to the future, although only a stopgap, in many ways. It was the vehicle that could tide the company over until the new car was launched, and on these two vehicles many dealers had staked their future. Despite the problems of the past they saw real potential

and the prospects of a line of good cars coming from the plant at Munich.

Late in 1959 a historic annual general meeting took place in Munich when the banks presented their plan to refinance BMW. To do this they would devalue the existing stock by half and issue new stocks to a group of banks and Daimler-Benz only. The stockholders were alarmed by this as their holdings would drop to half their value and the dealers more so after their years of work to build up their agencies. Together they forced an extension of the meeting and so the company was able to carry on.

Before the meeting could re-convene the 700 established itself as a car that sold well and was popular, while the big 500 series were dropped one by one.

In September 1961 the new medium-class car was shown at Frankfurt. It was the 1500 and a new BMW era and image was born. From it came a whole stream of desirable cars and at the end of 1963 BMW were able to issue a dividend to shareholders, something that had seemed impossible to consider four years earlier.

While this financial activity went on the motorcycles continued to be built, albeit in small numbers. For 1960 there were changes for the R60 which became the R60/2 with raised compression ratio, increased power output and stronger crankshaft and camshaft. At the front of the engine the fit for the dynamo rotor was also altered and improved. During the year both the R26 and R69 were replaced by new or revised models, while a sports 500, the R50S, also joined the line-up.

The new 250 cc model, the R27, was first seen with the rest of the range at the Frankfurt show in September 1960. In many ways it was the same as before but the compression ratio was increased to 8·2:1 and this helped the power up to 18 bhp at 7400 rpm, a useful increase on the R26. To remove the effects of vibration from the rider, the complete engine and gearbox unit, plus the exhaust pipe and silencer, were rubber mounted.

Above left **The R50S sports 500. This example has turn indicators built into the handlebar ends**
Left **The BMW for the 1960s, boxer engine, black finish, dualseat, smart, well made and pricey**

The R26 with all the cycle features of the twins except that the front brake is only a single leading shoe. Tank top toolbox

This was done with mounts at the front and rear of the crankcase and gearbox assembly, together with rubber buffers above and behind the cylinder head. Inside the engine a very neat automatic cam chain tensioner was fitted using a blade and a flat spring which also helped to damp out chain vibration.

The new engine was fitted into a slightly revised R26 frame, altered as needed to carry the new engine mountings but otherwise un-changed. Suspension, wheels, brakes, gearbox and transmission were all from the earlier model, as were the cycle fittings and switchgear mounted as always in the headlamp shell.

The new 600, the R69S, was a sports version of its more sober touring brother and was compli-mented by the sporting 500, the R50S. Both sports models had modified engines with a new timed rotary disc breather mounted on the front end of the camshaft and communicating with passages in the crankcase. The compression ratio for the 500 was 9·2:1 and the power output

35 bhp at 7650 rpm, while the figures for the larger twin were 9·5:1 and 42 at 7000. This was sufficient to push them along at 100 and 110 mph respectively under the right conditions while retaining the smoothness and silence of their predecessors.

For the chassis both sports twins used that already well proven on the earlier models and the only changes were to the gearbox ratios, which became a little closer, and to the steering damper. This became a hydraulic unit, while the friction type was retained by the touring twins. These ran on through 1961 with the fitting of the modified stronger crankshaft to the R50 together with the revised dynamo mounting already in use on the 600. Like this the type number was changed a little to R50/2. One new feature on all the models was the fitting of turn indicators in the handlebar ends in advance of some impend-ing German legislation. Neat but very prone to damage when the machine was being handled through a doorway.

The finish for the range continued to be in black with white lining, although on occasion this would be reversed to special order. The effect of the all-white paintwork was quite startling but still dignified and very much in the BMW image. Customers could also purchase accessories such as larger tank, dualseat, different handlebars, crash bars and prop-stand, along with other minor items, but all of which added to the already high price of the basic machine.

They may have been expensive motorcycles due to the high quality and small numbers built, but they brought their owners a very special form of travel and many special detail points. At that time unit construction was by no means universal and shaft drive a rarity. Then there was the adjustable load setting for the rear suspension, common enough by then but normally needing a spanner to set it while the BMW had its lever on each unit. The trail of the front fork could be altered for sidecar duty and there were

few machines left then with that facility. The use of 18 in. rims was not usual in the early 1960s, neither were alloy rims for models viewed more for touring than anything else.

Then there was the geared twistgrip. It was heavy, slow and suffered from backlash but gave a variable rate to throttle opening so aiding delicate control at tickover and fast response at speed. The combination of the bevel gear on the grip, the second gear in the cable box, the cam form attached to it, the chain wrapped round the cam and plenty of grease also gave a straight pull on the cable and hence a long life to the wire.

The front brake was a massive twin leading shoe design in a full-width hub with straight spokes, the air filter was also massive, the silencers large enough to reduce the exhaust

R26 on show at Geneva in 1958. Special finish with chrome plating on tank

noise to a whisper. As well as indicators it also had a headlamp main beam flasher.

The twins were machines for riding a long way in comfort. The combination of weight, its distribution, the suspension, the seat, the riding position, no vibration and quiet engine and exhaust made for a motorcycle which could be ridden fast for long periods. Plenty of other makes were faster over 50 miles but few could cover 500 as quickly or leave the rider as fresh at the end.

Not many could afford the BMW but enough did to keep it in production in the 1960s while the British industry began to crumble to pieces and a flood of small models came out of Japan to sell motorcycling, at first in the USA and then once again in Europe.

BMW continued with their well tried and trusted twins plus the single, although the shrinking motorcycle market shrank especially quicker for costly, well bred, refined models. It was the decade of the café racer, clip-ons and rearsets, exposed fork springs and many other features that were the antithesis of what the blue and white badge stood for. In that climate there were few changes to the machines. One that did take place was in 1963 when the two sports models had vibration dampers fitted to the front of the crankshaft, outboard of the dynamo. The damper was a steel ring sandwiched between rubber on its drive centre.

During that year the R50S was dropped, for customers either wanted the touring version or the larger and faster 600. Few bought the sports 500 so it went but the rest of the range continued year by year, to special order only in some countries.

Magazines continued to road test the black BMWs, and in 1963 *Motor Cycle* had an R50. They found it was good for 92 mph with powerful brakes and easy to handle, although the gearchange was still rather slow and the machine needed a good heave to get it on the centre stand. The controls were also rather heavy to

Above **Mammoth optional fuel tank beloved by many Continentals. Only BMW comfort allowed the rider to use the full contents in one sitting**

Top **A sports twin in white on show at Frankfurt. Very attractive**

Right **Radio aerial and rocker box lid protection bar have been added to this model**

operate but smooth in their action, while the many detail fittings were welcomed. A further test in *Motorcycle Sport* perhaps summed the model correctly as sophisticated with every detail well designed and beautifully made but, as an assembled motorcycle, giving a solid Teutonic feel lacking the grace, lightness and inspired simplicity of its competitors. It was unlikely that any of them were as well finished and most would be derelict when the BMW was still going about its business, but somehow it always looked and felt heavier than it really was.

A 1964 *Motor Cycling* road test of an R60 with a sidecar confirmed much of this and repeated the oft-made criticism of the gearchange, although they did find that it was much improved at speed. The outfit proved to be good for over 70 mph and the brakes on all three wheels worked well.

The weight of the BMW was a real factor in its performance and in the middle 1960s the factory was working on this using their ISDT model as a starting point. The prototype road model was based on this machine and had telescopic front forks, a much lighter duplex frame and a worked-on engine. The wheel hubs, rear fork and mudguards were also lighter, the last being in fibreglass and of sports section. Seat, tank and exhaust system were also changed to give a more sporting appearance and to shed more weight. The result was a loss of about 50 lb and, with the peppier engine, the prototype went very impressively. The machine was easier to flick from side to side through S-bends and even easier to get up on its stand. It was also not available and customers were to have to wait a few years before Munich were happy that they had a real improvement to offer.

For the home mechanic the BMW offered easy routine maintenance, although the points were not really as accessible as first thought, due to the presence of the front wheel. The big cover over the electrics could be awkward to remove as well. The worst point for many owners was the large, effective but expensive silencers. Thanks to the low running temperature of the well cooled

The final 250 cc BMW single, the R27 with rubber mounted engine, but otherwise much as the R26

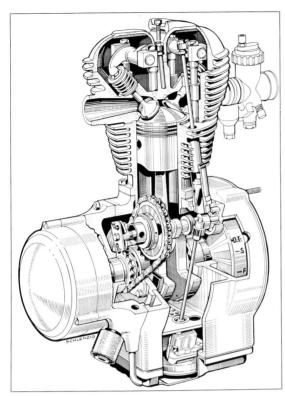

Line drawing of the R27 engine unit showing chain driven camshaft and typical valve rocker posts. Lengthy oil pump drive to submerge unit

engine it was hard to get these up to a decent heat to evaporate the corrosive condensate inside them. The result was rot from the inside and a short life. Some owners fitted stainless steel systems to overcome this as, although costly, they were cheaper in the long run for anyone keeping their machine for several years.

Many did do this, for BMW reliability had become a byword and remarkably few major problems occurred. Those that did were often due to sludge forming in the oil, to the detriment of the big end bearings and sometimes the back main. The latter had become a spherical roller bearing for the sports models as this allowed for the high engine speed and a modicum of deflection in the pressed up crankshaft.

It was that assembly that could cause trouble for its lubrication was based on the oil being flung out into pressed cups which led it into each crankpin and so to the rollers. It worked well with a low pressure pump as long as the oil and the system was clean. Any sludge formation at the wrong point could divert the oil flow with unfortunate results. The problem for owners was to decide if their particular machine, oil, way of driving, and climate produced sludge, and if so when to strip the engine to clean it out. In many cases only to find everything clean anyway.

The single cylinder R27 was not so prone to this problem due to the design of the single-throw crankshaft and because it ran hotter to the benefit of both oil and silencer. The 250 was usually driven hard for, with rubber mounting, none of the engine vibration got through to the rider and the combination of slow action throttle and very effective silencer would often disguise the speed at which the engine was turning over. The machine was in fact good for 80 mph with cruising all day at 70 normal. It was a very pleasant small capacity machine, although still with the odd BMW gearchange, but its price reached an unacceptable level in the end in most countries. By 1966 it was only available to special order in the UK and cost more than most 650 cc machines, so there were few customers.

In 1967 a change was made to the front suspension of the twins supplied to the United States with telescopic forks being fitted in place of the Earles type. These had always given a very good ride thanks in part to the multi-rate springs fitted to the units which allowed the forks to react to a matchstick and absorb the impact of running over a brick. Their snag was their weight when the forks had to be turned, for with weight comes inertia and this had its effect on the handling. In addition, they gave a feeling of remoteness to the steering that moved it away from total precision for the really fast and critical rider. For nearly all owners the Earles forks were fine and more comfortable just as the swing saddle gave a better ride than the dualseat.

Engine unit of the flat twin of the 1960s showing the very smooth lines of the complete unit

Combined ignition and lights key did not give much security. Rev-counter built into speedometer

During the year both the R27 and the R60/2 were dropped, which only left the touring 500 and sports 600 in their European form, with Earles forks and flat bars or USA style with teles and wide bars. Along the way 12 volt electrics became available, as did an uprated 6 volt dynamo prior to the 12 volt one. Right at the end an alternator appeared with rectifier and appropriate control unit, but as always the generator was mounted on the front end of the crankshaft.

Sales, however, dwindled for the machines

Design exercise of middle 1960s with light cycle parts to enhance performance. Even with old type engine it gave the machine a new sparkle

were expensive to make and costly to buy. There were not many private buyers about for motorcycles in the late 1960s and few chose the sober long lasting Munich twin. Preference went to the light if rather dated English and Italian machines with their excellent gearboxes and slick handling which offset the oil leaks and poor chrome plating respectively. If this failed to appeal then the new Japanese offerings would, with their very complete specifications, trouble-free electrics, and high power outputs. Their poor handling,·

mediocre paint finish and rock-hard tyres were anathema to the enthusiast but there were not too many of them left, the market was for machines with colour and style on the showroom floor.

BMW were dignified in appearance, still in their black finish with white lining. It was time for a change if they were to continue with motorcycles, for a few police bikes were never enough to keep the production line busy. So the machines had to change, with the marketing and company philosophy. During 1969 the old models were stopped to make way for the new, and the Friz concept took another twist—they moved the camshaft.

4 | New breed

During the 1960 period the Japanese motorcycle appeared as if from nowhere and sold itself world-wide. In 1958 few people, even motorcycle enthusiasts, knew anything about Honda, Suzuki or Yamaha but by the middle 1960s a mention of motorcycles to the layman brought an immediate response of 'Honda'.

The Japanese takeover of the industry did not start in 1960 but ten to fifteen years earlier when they founded their firms and learnt how to make motorcycles. Many of those early machines were crude with very low power outputs, many were glaring copies of European models, including BMW, but with an insatiable home market in the 1950s there was more than enough business to keep the better companies going. From that decade of struggle came the few remaining firms, by then large, well established and with a solid background of what not to do.

Using up-to-date marketing techniques along with competition programmes to highlight company achievements they moved into the export business and sold motorcycling to a world that had never considered two wheels before. Their products were colourful, well made, oil tight and had good electrics. Above all they were reliable, for even the smallest and cheapest was built to a high design and production standard.

The tank badge with the optional chrome plated panels offered for 1972. Not popular, not to the BMW image and not offered for 1973

No longer did cheap mean shoddy and poorly made.

At first only small machines up to 250 cc came out of the east and the UK makers selling in the USA were content as they continued to sell their usual large capacity twins in a traditional manner. They began to have some trouble when the new breed of motorcyclist moved up from his earlier 100 cc and 250 cc model to a 650 cc twin. It was normal practice for big to be thought better, but having bought his large twin the new owner expected the same standards from it as he had received from his small Japanese machines.

So Japan gave them to him and from this background came the Suzuki 500 twin, the Kawasaki triple, and the Honda 750 four. They brought a new dimension to motorcycling with features that previously had only been available to the racer or on a very few, costly road machines. The very fast Yamaha twins, along with the other makes of two-stroke engined machines, showed a level of power, consumption of fuel and narrowness of power band that had never been available before. The Honda four and the later Kawasaki 900 provided performance and sophistication to a level once only dreamed of. A new name was coined for these machines—super-bikes—and any maker who wanted to stay in business had to have a machine in that class.

For BMW to meet this challenge and retain their traditional image was not easy. They realised that it would be hard to meet the 125 mph top speed of the Honda or the acceleration of the triple. They also realised that their new machines would have to be marketed to the world, not just available for purchase as in the past. Fortunately much of this attitude aligned itself with the car division which had

Another light BMW, still with the old type engine and set up for off-road use. Important factory personnel in background

R75/5 on test leaves the line in 1971. 750 cc, drum brake
and old type headlamp with speedometer

taken on an up-market, executive style with a successful competition background on the race track.

Equally, the new models would be flat twins with shaft drive and would reflect the traditions of the Bavarian make in terms of silence, reliability, and rider's features. It was to be a machine for riding farther and faster than before but still retaining the comfort and handling that the Japanese machines could not offer.

Motorcycling was about to become respectable, and over the next decade the use of two wheels at all levels of society became accepted as traffic congestion worsened and petrol became more and more expensive. BMW found they were to be well placed for this expansion as they provided a machine for those who wanted something a little different, with an elite background, and still without any loss of reliability.

There were three new models announced in 1969 of 500, 600 and 750 cc capacity. All were very similar within the engine, as they used the same stroke, and identical externally as far as the cycle parts went. As expected, they were flat twins with unit gearbox and shaft drive. The frame was new and lighter, front suspension by very long travel telescopics, the electrics 12 volt, the petrol tank big, and starting by electric motor as standard for the two larger models.

The fundamental changes to the new engine were the crankshaft and the camshaft position beneath it. The pressed up assembly went to be replaced by a simple two-throw forging of very ample proportions, so it was more than adequate for the largest capacity and could cope with a further increase if desired. The one-piece crank meant split big ends and shell bearings, which in turn demanded a high pressure lubrication system. The rods and their shells came from one of the six cylinder cars where they survived a harder life, so had no problems in the twin. The Eaton oil pump was driven from the end of the camshaft and supplied the lubricant via a full flow filter.

The balance of a flat twin is excellent with only the rocking couple induced by the cylinder offset to mar perfection. To balance the different piston weights of the three capacities BMW devised small counterweights which were bolted to the crank web, the size of weight being selected to match the piston. The crankshaft turned in plain bearings front and rear with a pair of thrust washers either side of the latter to provide location. The bearings were in turn housed in a massive light alloy crankcase casting with detachable front support plate, and this casting stretched down to near the bottom of the sump and up to half cover the starter motor and the air cleaner.

On each side went a light alloy barrel with cast iron liner and on top of that went an alloy head. Four long through-studs held them in place with an extra short pair between each head and barrel. As was usual with BMW, the rocker supports were also held by the main head nuts and carried the rocker spindles between them. Each rocker had its adjustor at the inboard end and was enclosed by a one-piece alloy cover that internally separated the inlet and exhaust sides from one another. Each valve was held to its seat by a single spring retained by a collar and split cotters.

The pushrods were moved by well lightened cam followers that ran direct in the crankcase. The tubes that enclosed the pushrods were sealed to the case and into the barrel so acted as drains for the rocker box oil in addition to keeping the valve gear out of sight. The camshaft lay directly below the crankshaft on the engine centre-line and was driven from it by a duplex chain. This was kept in tension by a pad on a spring-loaded pivot arm. The camshaft ran directly in the rear crankcase wall and in a separate aluminium flange bearing at the front.

At the rear the camshaft drove the oil pump set in the crankcase wall and at the front it drove the rev-counter and carried the contact breaker cam and automatic advance mechanism. The drive chain was positioned immediately forward of the front main bearing and enclosed by a timing cover that carried a further outrigger main bearing for the crankshaft in the form of a ball race. The skew-drive gear for the rev-counter was mounted on the camshaft immediately in front of the chain sprocket and engaged a mating pinion carried in the timing cover. The cover also acted as a mounting for the contact points, ignition condenser, diode plate and alternator stator, the rotor being bolted to the front end of the crankshaft. All these electrical items were enclosed by a single aluminium cover that set off the front of the engine, and suitable seals kept the ignition side separate from the generator, with the wiring fed through grommets.

The underside of the crankcase was closed off by a shallow, finned sump complete with drain plug, and within it lay the oil pump pickup with strainer screen. The main oil filter lay across the crankcase low down near the front and from it oil passed via the front camshaft bearing to the front main, the rear main, the big ends, the cylinder walls and finally the valve gear. The ballrace main was lubricated by splash, courtesy of the cam chain dipping into the oil. An untimed disc breather was fitted high up at the rear of the crankcase and its outlet connected into the air filter chamber.

The air filter was mounted above the gearbox as of old, under a pair of aluminium covers. These connected it to curved intakes which led the clean air to the carburettor clipped to each cylinder head. These were slide-type Bing units for the two smaller models, but of the constant vacuum type for the 750. In all cases they were supplied with fuel via two taps with the pipes interconnected.

On the exhaust side finned nuts held the pipes to the ports and the exhausts were joined by a balance pipe just in front of the sump. Each pipe was clipped to a slightly uptilted silencer of generous size.

The top of the crankcase was finished off by another aluminium cover and this housed the electric starter and its solenoid. This engaged with a toothed ring on the flywheel to turn the engine and was of the pre-engaged type where the solenoid moved the pinion into mesh. It was driven via a slipping clutch and an electric safeguard prevented its engagement while the engine was running.

The three engine sizes were all based on a 70·6 mm stroke crankshaft and the bores were 67, 73·5 and 82 mm for the 500, 600 and 750 cc capacities. Power outputs were 32 bhp at 6400 rpm for the 500, 40 bhp at 6400 rpm for the 600 and 50 bhp at 6200 rpm for the 750. All three drove a single plate clutch with diaphragm spring bolted to the end of the crankshaft and operated by a lever at the rear of the gearbox. This was pulled by a cable and moved a pushrod in the gearbox input shaft.

All models used a four-speed gearbox of three-shaft type whose housing bolted to the back of the crankcase to complete the enclosure of the clutch. All the shafts turned in ball races, and the input one had a cam lobe shock absorber built into it. At its rear end went the kickstart ratchet and gear assembly which meshed with a gear set off to the left and coupled to the external kickstart pedal.

The input shaft drove a layshaft which in turn drove the output shaft stepped over to the right. Thus all the ratios were indirect and effected by a single gear pair. Gear selection was by two forks and done entirely on the output shaft by moving splined collars to engage with dogs on the free-running gears. All four gears on the layshaft were fixed to it and revolved with it. The selector forks were moved by a face camplate itself turned by a positive stop mechanism.

The gearbox shell was of the barrel type with a front wall and open rear closed by a separate casting. This housed the speedometer drive taken from the output shaft, and the latter carried an output flange at its rear end. This

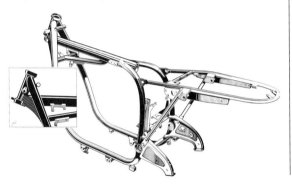

Above **Both sides of the R75/5 cylinder head. Simple two valve, hemispherical design**

Left **One piece crankshaft with plain bearings. Different balance weights fitted to suit various engine sizes**

Below **Frame of the /5 series with bolted on sub-frame**

Bob Lutz, then sales director of BMW Munich outside the works on a 1972 model. Now with Ford

bolted to a mating flange with a Hooke-type universal joint positioned to lay on the pivot point of the swinging fork. The drive shaft ran in the right fork leg and connected to a gear form coupling that took the drive to the bevel box pinion. This coupling accommodated any slight misalignment and could also deal with any movement or flexure of the frame or fork arm.

The bevel box was of the normal BMW type with spiral bevel gears, although strengthened from earlier models. Due to the layout of the gears the torque reaction as the drive was applied caused the rear suspension to lift, and in practice this balanced the usual acceleration loads to give an anti-squat feature. On first acquaintance it felt odd.

The new engine and transmission went into a new frame a useful amount lighter than the old. It was built up using oval section tubing joined by an inert gas welding process that produced very strong, neat welds when done properly. At BMW it was. The frame was in two parts with the main loop based on a single large diameter top tube welded to the bottom of the headstock. At the

rear, two tubes were welded each splayed out to run down to the rear fork pivot and then down to turn forward under the engine unit. Each turned up and then in to meet the top of the headstock and form the two major loops. The down tubes were cross-braced just below the steering head and a further tube ran back from this brace to the underside of the top tube. Substantial gussets ensured that the headstock remained true to the frame, which was cross-braced between the loops above and below the fork pivot. From this area two steel pressings ran back to support the silencers and pillion footrests.

The subframe was separate and bolted on at four points. It comprised a simple loop that ran back to support the seat and rear mudguard with a strut welded to each side. Brackets for the tops of the suspension units were welded to the tube junction and subsidiary lugs added, including those for the tyre pump now tucked out of sight under the seat.

The rear fork was built up and pivoted as before on two pins and a pair of taper roller bearings. The rear bevel box bolted to the right fork arm and carried the damper mounting. The left fork arm was split at the end to clamp on the rear wheel spindle. It was also chamfered so that as the spindle was turned with a tommy bar it was levered out from the wheel.

The rear spring units retained the old type load adjusters with short levers but exposed the springs by omitting the lower spring covers. To compensate they were either plated or painted. At the front went telescopic forks with well over 200 mm of travel, pivoting on taper roller bearings and with the springs enclosed in handsome rubber gaiters. Two-way hydraulic damping was used and, with progressive action

Left above **Frontal aspect of the R75/5 showing the fork gaiters and indicators**

Left **Handlebars, controls, tank and instruments of the 750 cc /5 model**

57

springs, kept a firm control over the large fork movement. To keep the steering itself under control a friction damper was fitted with a hydraulic one available as an option.

The wheels were typical BMW with alloy rims and straight spokes connecting to alloy hubs with a cast steel drum surface. Both brakes were the same 200 mm diameter as before but the width was cut down to 30 mm to counteract the effect of a new and improved lining material. It had been developed for Porsche and the racing BMW saloon car and worked a little too well in the old width. The hubs were finished off with a spinning to close off the casting recesses on the non-brake side. The front brake had twin leading shoes operated by the cable outer and inner as always, and anchored by a torque arm to the right fork leg. At the rear a single leading shoe was fitted and the cam lever was pulled by a rod linked to the right brake pedal.

The tyres were of different sizes with a 3·25 × 19 in. at the front and 4·00 × 18 in. at the rear. Each was shrouded by a reinforced poly-ester mudguard, the front unsprung and sup-ported at the forks and by a rear stay, while the rear one was deeper, sprung and run up under the dualseat. This was large enough for two, com-fortable and fitted with a small grab handle on each side at the rear. The whole seat pivoted on its right side and lifting it gave access to the tool tray, and beneath it the large 12 volt battery. The tool kit continued in the BMW tradition with an extensive range of tools neatly packed in a tool roll, along with a cleaning cloth with BMW logo.

A 22-litre tank was fitted with a central filler, twin taps and deep sides that reached down to the engine. It had a kneegrip on each side and was held by two wing nuts at the rear. Below the engine on the lower frame tubes went a centre stand carefully sited at the centre balance point of the machine so that either wheel could easily be removed without the need of a jack, housebrick or block of wood to keep the model on an even keel. Further forward a prop-stand

BMW in use in London. Avon fairing complements the machine. Metal rear indicator shells show

was provided. The centre stand had its unusual aspects for the spring centres were such that it would stay down when pushed to the ground while the rider moved his foot onto the left leg to force the machine up on the stand. It did the same trick coming off so you had to remember to flick the stand up after rolling the bike off it. All this activity was aided by a good sized handle bolted to the left of the machine just below the seat at the front.

On the electrical side it was much as before despite the change in voltage. The ignition and lights were still turned on by the Bosch key used by so many German machines. This was pushed in for ignition and then turned to left or right for

By way of contrast a /6 model of the 600. In fact one of the nicest, if not the fastest, of the more modern BMWs

the lights. To this was added a combined starter and indicator switch on the right handlebar, and dip, headlamp flasher and horn on the left. Warning lights were provided for oil pressure, main beam, neutral, charge and indicators, while the headlamp shell carried a combined speedometer and rev-counter. This last was really too small to be of any real assistance. The turn indicators were mounted from the lower fork crown at the front and either side of the rear lamp at that end. They were of a substantial size. The stop light had the benefit of being switched on by use of either brake control, then a less than common feature, and models destined for the USA had side reflectors fitted to the turn signal

bodies as well as high and very wide handlebars. The handlebar controls came from the earlier model and the bars sprouted mirrors, a feature that had become a real need as traffic grew more dense.

The appearance of the complete machine was a little odd. The older machines with Earles forks had always conveyed the impression of length and solid, worthwhile weight. The new ones could be forgiven for the thought that they had recently suffered a minor accident.

This arose from a combination of factors. First, the whole engine unit was one solid mass that ran right up to the fuel tank. It may have been alloy and full of air in many places, but it was short and high with the sump clearly visible below the frame tubes. Secondly, the machine as a whole was shorter and more compact than in the past with a considerably higher seat. Finally, there was the combined line of the engine and the silencers. The first was canted back for the sound technical reason of lining up the gearbox output shaft and the drive shaft in the fork leg to minimise angular movement in the universal joint. By keeping this down it would transmit more power at a higher speed in an engineering trade-off. The silencers on the other hand were tilted up at their ends, which was beneficial in that condensate would tend to flow down to the hotter exhaust pipes and be evaporated to the good health and long life of the system. Unfortunately the combination of these aspects gave the machine a slightly broken-back look just as if it had been in a small shunt and had bent the frame slightly but not the forks.

R50/5 with single mirror on left only. The cycle parts were common for all three capacities as was much of the engine

Top of the /5 series, the 750 cc model. Note kick-start, grab rails, knee pads; all will change

Reaction was fairly mixed and further complicated by a novel feature for a BMW motorcycle—colour. The mudguards and tank were still lined as stylishly as always but those two items were available in silver grey on the 500 and 750, while the 600 remained in the traditional black. Frames, fork tops, headlamps and minor cycle parts continued in black so the colour made little difference on the production line. Before long other colours were to become available as it was found that buyers would put their money down for a BMW with red or green petrol tank and mudguards.

There was a further complaint from sidecar men as the new frame was not designed to withstand the stresses of chair use and attaching one made the guarantee void. This failed to disturb the more determined and it was not too long before owners found ways of coupling their chairs to the new flat twin.

The new range of motorcycles also had a new factory in which to be built. The expansion of the car business had forced the company to concentrate its men and factory space in and around Munich on this side and this left no room for the two-wheelers. The new factory was set up at Spandau in West Berlin and within a few years was fully tooled to build the complete machine and all its parts. Much of the labour force came from the Balkans, with Yugoslavs predominating, for German labour was hard to come by in that town. It made no difference for, once trained, the immigrants proved to be very good, very thorough, and keen to work overtime to meet

Middle model, the R60/5 which many riders preferred for its combination of smoothness and urge. Well used this one

the many demands for the new models.

Quality control was and is to a very high standard with many pieces of special equipment to ensure this. Along with these and the special tools and machines were the skilled welders joining frame tubes, and the lining ladies with just brush, finger and sure eye to apply the finishing touches to mudguards and petrol tanks.

The big question was how did the new models perform on the road, and the answer came back—as BMWs. There were differences for the ground clearance was increased with the raised barrels and canted silencers, engine noise was up a touch thanks to the alloy cylinders and the pundits, as usual, said the finish was worse. What

was unchanged was the basic BMW ability to cover a big distance quickly, quietly and without straining man or machine. The big petrol tank helped this as it allowed 200 miles to be covered non-stop, while the seat, the riding position, the controls and the whole as-one feel completed this.

A host of touring extras could be added but while useful none were essential to that basic style of the machine, they merely complimented it. The road tests that followed brought much of this out and, while some writers suggested that it was slow for its size, others enthused over its now relatively light weight and ease of handling. It was not that the new models had lost much weight, although they were down despite the added electric starter, it was the rest of the world that had got heavier. No longer was the twin looked on as that heavy Germanic bike.

Above **Works jig for assembly of the frame by immigrant worker**

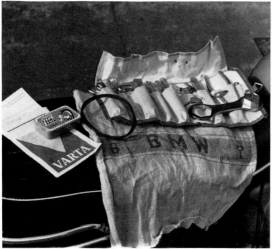

The new models were comfortable thanks in part to the long travel front forks, but these also played their part in a handling problem that arose in some cases. The soft forks combined with the short wheelbase and the abrupt effect of the nearly rigid drive line all made for substantial changes in the machine's attitude on occasion. This could have a severe effect on the fork

Left **The famous BMW toolkit and its hand towel with logo. All you need and so seldom required**

Factory assembly line with cylinder head being torqued down by air driven sockets

geometry and in turn this was allied to a frame that had a degree of compliance fore and aft. This in turn demanded just the right combination of wheel rim truth and tyre construction. Put together it added up to a machine with quick accurate steering that was more than good enough for most owners. A few, however, ran into problems. Either they rode harder, or they had the wide US bars, or they failed to keep to the original fit of Metzeler or Continental tyres. The result could be a high speed weave of very unsettling proportions even with the steering

damper in use. Either way that item was not to be thought of as an ornament.

That aside, the new 'slash 5' or /5 machines made satisfactory motorcycles especially for the people who could put down the not inconsiderable amount of money needed to buy one. Owners of BMW had always had to be defensive as to their machine's price but for those who wanted that form of motorcycling it was not thought too expensive as long as they could afford it.

There were one or two minor problems. The gearchange was still a special art requiring six months practice and a Bavarian sacrifice at midnight, some said. The carburettors on the 750 were so large that they interfered with a normal

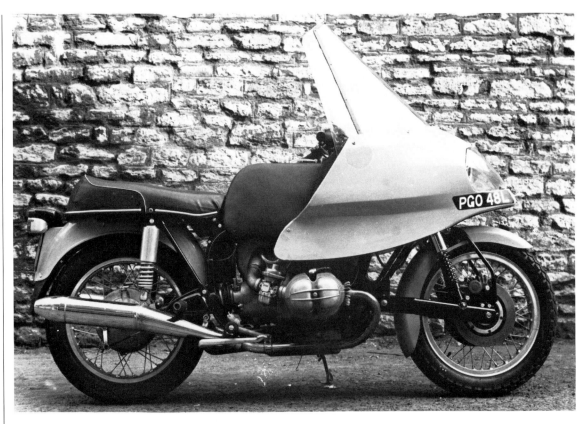

Special tank, special fairing and very special Difazio front suspension with hub centre steering

rider's legs, although those with large feet had no problems. The horn worked well but the electrical handlebar switches were neither ergonomic or too reliable. The mudguards were rather inadequate but the controls were good, although the choke lever, bolted to the left side of the crankcase high up, took some getting used to. The two petrol taps were positive in action and easy to use. The exhaust pipes turned blue at the front bend in the first 1000 miles.

With all these points went a 110 mph top speed for the 750 at a time when plenty of other machines of that size were good for 120 plus, and an appreciation of how special the BMW was in 1970 begins to emerge. It was necessary to ride one a long way to find this out and the reasons why they were bought despite the high price.

BMW left the new range alone for a year or

two as they built up production but did introduce some changes for 1972. Just as owners were beginning to come to terms with the new appearance they changed it by adding chrome plated panels on each side of the tank and similarly finished small side panels to hide the battery. The tank size was also reduced although the big one was still available, while the rear wheel rim went up to a WM3. The propstand was modified so that it automatically retracted, and this was not too successful as it was hidden by the left cylinder and it was all too easy to ease the machine up, not realise the stand had sprung up, and then lay the model on the left rocker cover.

A revised dualseat was fitted and was bigger,

Very nice R75/5 with Watsonian sidecar and extra wide handlebars

nothing to spoil the image of the machine. The items that did do this were the battery panels which, with their plating and false air slots, presented a very garish picture to the world. The chrome tank panels did nothing to help this and were further spoiled by a trio of styling bars each side of the traditional round badge. The new tank shape was perhaps less attractive than the deeper and larger one and again the panels, when fitted, did nothing to enhance the looks. BMWs always wore their deep, ample capacity tanks with a superior air and just did not look right with anything less.

For 1973 most of the machines imported into the UK were fitted with the bigger tank, although by then the chrome panels had been dropped along with the side panels. That year the colour choice was red or metallic green for the tank and mudguards with white lining in either case.

During the year a significant change was made to cure the handling problems suffered by a few riders. To deal with these the swinging fork was lengthened some 50 mm and also stiffened up. This increased the wheel-base as well and curtailed the rather abrupt change of attitude that slamming the throttle shut and the front brake hard on previously provoked.

That year was also the motorcycle company's Golden Jubilee, for it was 50 years since the R32 had made its debut. To celebrate this the UK concessionaires arranged for the Berlin factory to produce a special R75/5 in a golden paint. They then contrived to take advantage of the 'L' suffix used for 1973 registrations to get it the number BMW 50L, the suffix letter being of course the Roman numeral for 50. The machine was shown at a London exhibition in January of that year and was fitted with a transistor radio made by Blauplunkt and connected to tiny speakers in the rider's helmet. The machine was insured for £10,000.

On that note the /5 series bowed out late in 1973 to make way for a revised range that was very similar but with some useful changes.

firmer, ribbed along its length, and fitted with a most useful handrail that ran round the back. With it went rear suspension units with stiffer damping and the overall result was a firmer ride, still very comfortable.

Internally, the flywheel weight went down by a kilo and this was especially noticeable on the 500 which had less pure grunt to overcome engine inertia. At the rear of the machine the bevel box ratio was changed on the 750 to lower the gearing.

More colours appeared, with black, white or red being normal and metallics in silver, blue, green or mustard being available. These met with a mixed reaction from owners but in fact did

5 | Modern Times

Fitting the longer rear fork and increasing the wheelbase overcame the handling problems of the /5 models, but these still had an old-fashioned air in one or two places which did not sit easily with the styling of the rest of the machine. It was the headlamp and the instruments that spoilt the picture and this was one of the areas that BMW chose to improve on the new models introduced late in 1973.

These were the /6 machines and they were four in number based heavily on their predecessors. The smallest twin, the 500, was dropped but as compensation two models appeared at the other end of the scale, one a sports version with cockpit fairing and a new style of finish. The 600 and 750 models continued using the same engine dimensions as before and the new model, a 900, was a further development using the same crankshaft. To do this the bore was taken out to 90 mm so the unit was well over-square and the pistons were becoming a considerable weight.

Along with the new larger engine there was a new five-speed gearbox for all models and disc brakes as standard on all except the 600. The headlamp, instruments, side panels, silencers and control levers were all altered and all played their part in bringing about the more modern image of the range.

Within the engine itself there was a strengthening of the crankcase to deal with the increased power and torque, and for the sports 900 a warmer cam form. With it went a

An early 600 cc example of the /6 series with extended
wheelbase and revised headlamp and instrument
console. This is one of the nicest modern style BMWs

compression ratio of 9·5:1 as distinct from the 9·0:1 which sufficed for the standard model, and 38 mm Dell'Orto carburettors with accelerator pumps in place of the 32 mm Bings also fitted to the 750. At the front of all the crankshafts went an alternator with more windings and another 100 watts of output to make 280 in all, and this was supplied to a much bigger battery of 25 ampere-hours. This replaced the old 15 ah type and ensured that there was plenty of electrical power available for turning the largest engine over on the coldest morning.

Otherwise the engines were as before, aside from the badges on each crankcase flank above the barrels. They continued to drive a single-plate clutch but this did have a heavier clutch spring and reinforced pressure ring. That in turn drove a new feature for a road model BMW, a five-speed gearbox. This was based on the three-shaft type already in use and the extra gear was selected by sliding the centre gear on the input shaft. All models had the same box and internal ratios and these were modified so that the four lower ones were all lowered but remained in the same relationship to one another. Fifth remained the same as the old top so the overall effect was to lower bottom and provide a useful fourth. The other features of the gearbox such as the input shaft shock absorber and the transverse kickstart meshing with it remained.

The gearchange mechanism was completely revised and had two camplates geared together to move the selector forks, one moving the input shaft gear for third and fourth, while the two others moved output shaft gears to select first and second on one selector fork, and fifth by itself on the other. While the gearbox layout would improve the gearchange it seemed that BMW had still not taken heed of gear inertia and the need to move two gears, one into mesh and the other out, on each change. Only from 2nd to 3rd or 4th to 5th did this happen.

The new assembly of gears and mechanism was fitted into a new shell which closely followed the lines of the old with an end cover at the rear,

The R75/6 with single front disc brake. This always stops the machine but feels borderline. New silencers

Left **The R90/6 produced by boring the engine out further still. Same cycle parts as the 750. Dog leg levers plus new grab rail. 'Badge' on engine now blackened**

ball races for all three shafts, and air cleaner housing on top of it. The output flange remained in the same place and drove the same shaft to the rear bevel box. A variety of axle ratios were available ranging from the 3·36 of the 600 to the 3·00 of the /S and further on to its option of 2·91. The lowest ratio was the 600 option at 3·56.

The frames of the /6 'Bee-Emm' models were very much as used for the last of the /5 machines with the longer wheelbase. For the 900 they were reinforced at various points and all retained the forks and rear units from the earlier machines. A hydraulic steering damper was fitted as standard with an ingenious setting arrangement. The damper knob was marked 0, 1 and 2 and turning it the 90 degrees between each marking moved the front end of the damper one stage away from the centre of the steering column. Thus on '0' it was at the same centre so had no effect while position '2' gave it twice the leverage it had at '1'. It worked very well and even in town it was hard to detect any roll using position '2'. Out of town it kept a firm control on the front end.

The wheels continued with the straight spokes and at the rear the 200 mm diameter drum brakes on all models. At the front there was a change and some variety. As standard the 600 continued with the twin leading shoe front brake but the 750 and touring 900 had a single disc, while the sports 900 came with twin discs. In addition all the other models had the option of the twin discs and the 600 that of the single unit.

The discs were hydraulically operated and each caliper mounted on a hinge pin parallel to the fork tube axis so free to swing about the disc. The calipers, unique to BMW, went between lugs on the fork behind the leg to keep their weight close to the steering axis and, unlike other makers, BMW chose to mount the master cylinder on the top frame tube under the tank and connect it by cable to the front brake lever. Thus it was out of reach of any passerby but kept the rider informed as to its level of oil with a float switch and warning light.

Above **Top of the range was the R90S with cockpit fairing. Twin front discs, not drilled on the early models**

Left **Powerhouse of the R90S. Dell 'Orto carburettors with accelerator pumps, drilled front discs**

A minor change for the drum brake at the rear, and when fitted at the front, was the removal of the shiny backplate, the back of the drum hub and its webs being left open to view. Small holes were cast in the hub between pairs of webs to allow the brake linings to be checked for wear without dismantling the wheel.

Much of the improved appearance of the new models came from the new headlamp shell with its 180 mm Bosch quartz-iodine light unit. Above it was a neat console containing the speedometer and rev-counter side by side with a vertical row of warning lights between them. These were, from the top, brake fluid, neutral, battery charge, oil pressure and turn indicators with the main headlamp beam warning light set in the rev-counter on the right. The old ignition and light switch went and was replaced by one which was also the left headlamp shell fixing. It was hardly the most secure device as it was easy enough to open the lamp shell and join the switch wires together. A neater touch was the wiring to the starter solenoid as this took in both the gearbox neutral switch and the clutch lever.

If in neutral the starter would work normally but, if in gear, the clutch had to be pulled in before current would reach the motor.

The appearance of the standard models was assisted by a change to the silencers which were no longer bent so drastically halfway along their length. The side panels returned without any adornment and were coloured to match the tank and mudguards. In their new form they added a touch of dignity to the machine, and among the minor details were matt black control levers, a helmet lock, adjustable pillion rests and an optional 22-litre fuel tank.

Colours for the standard models were black, white, blue, green, red, curry and polaris all with the traditional twin lines on tank and guards.

Then there was the S model, the sports 900, with a style and finish all of its own. As far as the mechanics were concerned it was a case of more power, higher gearing and twin discs to stop it, but these were not the reasons that it was stared at. It came with a neat and stylish 'bikini' cockpit fairing and bubble screen in which were mounted a voltmeter and that most useful motorcycle accessory, a clock. It even had a sweep second-hand and could be read with ease at 100 mph.

There was a graceful 24-litre fuel tank with flowing lines, and behind it a wider dualseat with tail all of which hinged up as usual to expose, as always, the sealed but removable tool tray, the push-on BMW tyre pump and, behind the seat, a rear storage compartment. It was a style that was

Fast guard dog. No one is going to mess with this R90S. New black plastic indicators now

Cockpit of the R90S with normal /6 instruments plus voltmeter and useful clock. Styled switches were later re-arranged

BMW petrol tap full of Teutonic logic. Behind it is the inaccessible oil dipstick and filler hole

to become popular over the years.

The side panels carried the machine size '900 cc' and the engine the machine type, BMW R90S. At the front the fork gaiters were left off, a rare concession to styling. The finish was given by BMW as 'silver applied by hand' and this indicated the difficulty in describing something that was totally removed from a conventional coat of paint. It was, in a way, a custom finish but one as carried out by Bavarians for discerning motorcyclists. The colour was a silver but it varied from this to a deep smoked grey nearly reaching black. The area around the headlamp and the tank sides were perhaps the lightest in colour with a gradual change to grey at the edges, which were also lined in gold. This colour and the line extended to the seat base and tail which carried the BMW roundel and the legend '90S'. It made for a very handsome and very striking motorcycle.

BMW gave the new range a good launch in

Munich quickly followed up with plenty of test rides and much reporting in the press. They were offering a prestige product and meant the world of two wheels to know it and to have plenty of opportunity to read about the machines. To this end the UK test fleet alone soon amounted to 18 machines so that, whether one was wanted for a full test, a stand-in or background to another product advert, there was a flat twin available, willing and able.

And tested they were by all the specialist magazines who quickly found that the 90 mph claimed for the 600 with the rider sitting up, and the 125 mph for the /S with rider prone, were both correct. The claimed consumption figures were all taken at 68 mph and lay between 55 and 63 mpg, a very fair figure for cruising at a reasonable gait.

Apart from adding the capacity to the side panels during 1974 BMW left the machines alone that year, but for 1975 there were a number of changes. At the front end the forks were strengthened and the front wheel spindle went up from 14 mm to 17 mm. The engine unit lost the kickstart pedal although this was still available as an option. It was not an easy one to fit as a new gearbox end cover had to be used along with the pedal and gears. The need for it was made less necessary by an increase in the starter motor power by 20 per cent to 0·6 hp. This made it even more certain that the biggest engine would turn over on a cold morning.

More noticeable were the new electrical switches on the handlebars, which were mounted into the parts which carried the controls, twistgrip and mirrors. The two sides were contrived to look the same, although the switches had totally different functions and movements and these functions continued, to an extent, to be in the same places as before. Each housing carried three switches and those on the left were for the horn and lights. Nearest the rider's hand, and thus furthest outboard, was

Skilled lady lining the tank with brush, finger and experience

the horn button of square shape that was just pressed in as required. Below it was a rocker switch that provided the main and dip beams as it clicked up and down, and also could be pressed further down against a spring to flash the main beam. It did this regardless of whether the lights were switched on or not and was spring returned.

Inboard of the horn button was the lights switch which was rotary with a small ear on its knob for the thumb or finger to work and also to indicate the switch position which could be off, pilot or head. It was yellow and balanced on the right by a similar style red lever acting as the killswitch with an 'off' position either side of the run one. Outboard of that was the starter button which matched the horn one, and below it the turn indicator switch. The control for this matched the dip and flash one on the left and all the rider had to remember was up for left and down for right. Logical, but none too easy to manage especially on the /S at roundabouts in the wet.

Another noticeable change was to the disc brakes which were drilled to assist wet weather braking, a very real problem at that time and one not fully solved before the advent of sintered pads in the late 1970s. On the BMW the hole pattern gave rise to a whistling sound when the brake was applied and this at first seemed odd and then reassuring.

A useful change and improvement was made to the petrol taps, not before time as the old type were prone to leaking. The new ones set a standard and style that was taken up by many other makes. The lever was big enough for a gloved hand to get hold of, and both taps could be seen from the saddle when riding. Positions were down for on, up for reserve on and—as they looked odd—start searching for petrol, and sideways off whether they pointed fore or aft.

Right **R100/7 well equipped for touring with high bars, screen, panniers, crash bars and extra lights, all from the factory**

They also had a nice positive feel and clicked into position.

There were also minor changes in the casting finish which became smoother and in the clutch lever ratio which was altered to make the action lighter. The sports 900 was made available in an alternative colour of Daytona orange with red lining in the same style as the silver smoke, which continued to be available. For the standard models the range of colours was continued.

For 1976 there was little visible change but internally a good many minor improvements. One of the few external points was the addition of a rib to the cylinder finning to prevent ringing. Less obvious was the fact that the barrels themselves were new, wider and no longer had a base gasket, a sealant being used instead. Underneath a deeper oil sump was fitted, not to increase the capacity but to lower the oil level away from the crankshaft.

The camshaft front bearing was increased in size from 12 mm to 20 mm and its seal by an equivalent amount. The pushrods changed their material to aluminium which matched the barrel better, while the rocker arms were modified for improved stiffness. There were detail changes to the cylinder head, the starter gears and to the Dell'Orto carburettors on the sports 900.

On the chassis side the front fork yokes were made wider to enable the latest tyres to be fitted and the damping modified to make the ride more positive. At the rear the dampers were used in matched pairs and the rear fork was strengthened. The brake master cylinder and calipers were both increased in size, and in the UK the price was unchanged.

On the road this added up to four very individual machines which consistently were rated overall top for motorcycling, and the smallest model, the 600, was perhaps the

Wind tunnel testing the R100RS to obtain the right aerodynamic effects for rider protection and machine stability

78

The R100RS during a road test in 1976. Single seat and wire spoke wheels

sweetest. All the flat twins had certain characteristics to a greater or lesser degree. All turned their exhaust pipes blue. If the throttle was blipped with the machine in neutral and not on the centre stand it laid over to the right. On shutting off it laid left. As the clutch bit the drive pinion in the rear axle would try to run round the crown wheel and so lift the rear of the machine and the seat. Tight manoeuvring with the cylinders stuck out each side and the lift of the seat each time the clutch bit took some getting used to.

Then there was the shake of the engine unit. Not really a vibration more the effect of the two pistons banging in and out in a light frame. The result was a quiver and shudder at tickover as the whole unit seemed to shake itself, and this was more pronounced with the 900 than the 600. The shudder continued as the machine moved off and both mirrors were a blur until the engine

reached 3000 rpm. Then it all stopped and stayed smoothed out all the way to the red line. A nuisance in town but once away in the country so smooth.

The 600 was sweet and amiable and most suitable for pottering in lanes for an hour or two at 1500 rpm. Then it could turn out onto a main road and the counter would run straight round to the red line without pause or hesitation. It accelerated no better than a contemporary 350 cc two-stroke twin but did this in a totally different manner. It was without quirk when running but the spring-back propstand could catch one out. Take two paces and look back to find the bike on the verge of toppling over. The

Front end of the R100RS showing the optional cast alloy wheels and the front spoiler

Right side minor controls. The turn indicator was awkward to use in traffic and was later moved

centre stand was no great problem to use.

The handling was light but sure even over the bumps of the TT course, and if the front wheel were to leave the ground at 80 mph the steering damper kept matters well in hand. It also stopped well, but most of all it was comfortable. It was a machine that a rider could just keep riding without strain for over 200 miles until the need arose for petrol.

The 750 was the same but a little more so, and the 900 even more, although the sweetness of the smaller engines was beginning to fade. They did everything the 600 did—a little faster perhaps but also a touch less pleasantly. Only with the front brake was there a noticeable

change for the single disc was good but not good enough to give total confidence. It would always stop the machine and a solo rider but gave that indefinable feeling that with two-up it would not be able to fully cope and braking distances would have to be revised. It never faded but always felt to be on the point of doing so.

The sports 900 front brake was very different. From any pace it just stopped the machine dead with an air that the addition of a passenger, top box, panniers and a ton of luggage would make no difference. Not that one loaded a /S that much for, aside from the all up weight limit of 398 kg, it was a machine to ride solo.

The S was a very, very fast machine over any

The R80/7 which replaced the 750 cc model. A very nice motorcycle for riding far and fast

sort of road and was perhaps the machine which firmly cemented the 'modern' BMW. It was also comfortable. It was the one model where the in-line crankshaft did produce an effect on the handling but not the gyroscopic one much talked of. It was simply that when one changed down from third to second one could feel the machine lean right and then left. At higher speeds the machine's inertia held it true but that one change done under heavy braking did make it wriggle. After the third time it was ignored.

That aside, the gearchange was BMW, good but by no means as quick and slick as the better Japanese or the knife-into-butter of the Norton. It was adequate and fast enough for the machine but nothing to enthuse over.

The S accelerated very well and was most satisfying to take away from the traffic lights for, as the clutch took up, the seat would rise and then squat as the power took hold so the bike seemed to huddle into the ground and just go. Third was good for over 80 mph and fourth for 100 mph.

About that time BMW in the UK took their sales in a new direction. They had built up from 200 units in 1972 to 1000 in 1975 which, while small compared with the 10,000 plus in the States, was an impressive improvement. To augment this they began to market a machine modified for police use and, in 1976, had 50 on duty with four forces. Within a year they had increased deliveries tenfold to 500 machines, and

Above **R100RS set new standards in integral rider protection**

Below **Revised gearchange linkage that moved the pedal pivot back. Ball joints connect**

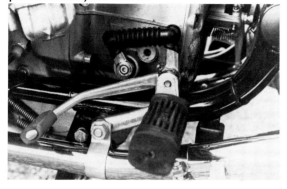

passed the 1000 total in May 1978. That machine was delivered by helicopter to a police function at Thruxton and went to the Hampshire force. By 1980 some 46 forces were using BMW machines and on 22 August 1980 the 2000th model arrived at Earls Court for press day of the Motorcycle Show. It had been ridden from Munich by Neale Shilton, the head of the BMW Police department and the sales of the special models and the cars were due to his expertise. He had been in the industry for 34 years and during many had been involved with police sales, first with Triumph, then Norton, and finally BMW.

During 1976 the existing four models were discontinued and five new or modified ones took their place. This was very much in the BMW tradition of gradual development and normally would have produced no surprises. Three of the models had larger engines still based on the same crankshaft so were even more over-square than the 900. The feature that grabbed the attention was a fairing for the top model but not one that was added on as an afterthought. For BMW that was not good enough, so they developed their fairing in a wind tunnel and the result was a machine that jumped ahead of the field for several years in one move.

The new range was given the type number /7 and came in 600, 750 and 1000 cc sizes. To achieve the largest size meant boring the engine out even further to 94 mm and, as with the 900, the sports engine ran on a slightly higher compression ratio than the standard unit. The remaining two smaller engines kept to their existing dimensions. From the outside only the crankcase badges and new style rocker boxes gave any indication of the alterations, but internally there were a number.

The breathing system was changed and the barrels sealed to the crankcase with 'O' rings to make quite certain there was no oil leakage. The barrel fins were made shorter but thicker to improve heat flow and reduce fin vibration. The valve gear was modified with alloy tappets and detail attention to the rockers and valve springs. The new rocker covers had a distinctive angular shape with four horizontal ribs and were either left natural or coloured black with the ribs faced off to shine. The largest engine was given larger Bing carburettors and the faired model also had slightly bigger exhaust pipes. Cold starting was made even more certain by changing the gearing from starter to flywheel.

Above the engine on all models went the 24-litre tank fitted with a flush filler cap with lift-up handle for unscrewing and lock to keep the contents safe. The frame was further improved by additional bracing around the headstock and the rear units lost their upper spring covers. They retained the short levers for easy setting to suit their load. The rear brake remained as before but the drum at the front of the 600 was replaced by a single disc also used by the 750 and standard 1000. The two sports 1000s had twin discs and all models still retained wire wheels with straight spokes and alloy rims. For the faired model cast alloy wheels with an intricate network of spokes were available as an option.

It was the faired model, the R100RS, that caused the sensation. For the first time a motorcycle was being offered for sale with an integrated fairing without the option. It was of the dolphin type with a screen that turned up at the top to tip the air up and over the rider and perhaps just leave him with a faint draught behind his head at 80 mph. A spoiler ran along each side to aid downforce on the front wheel, the sides went right round the barrels and a grille went before the engine. The turn signals were set flush into the fairing and the mirrors styled to it. The headlamp was set back into the fairing nose and sealed off with a sheet of approved glass set in flush to avoid any sharp edges. It was finished in German racing silver and was an immediate sensation.

It was the only model that was still fitted with a steering damper as standard, although this was an option for the others. Despite the longer, more heavily braced frame there was still the occasional bike and rider who needed it. The most irritating feature of the fairing was the bellows that sealed each fork leg to it as these proved to have a disconcerting habit of dropping down the fork leg to look like a schoolboy with a sock at half-mast. It was not dignified and at first proved to be a real snag as the bellows defied all attempts to restrain them from obeying the call of gravity.

Inside the fairing went the clock and voltmeter from the S model with the ignition key between

R65 on test. One of the two machines introduced to fill a gap at the lower end of the capacity scale

them. Both the dipswitch and the turn indicator switch had grown an ear to aid their use and the faired model had the choice of two seats, each with the faired tail that was part of the aerodynamics. The standard seat was really for solo use and very occasional and slim passengers, but the other took two in comfort.

BMW had moved on again for the serious motorcyclist for the RS could cruise at 100 mph plus without tiring the rider and only calling for fuel stops every 200 miles or so. They managed this with the rider neatly tucked in behind the screen and fairing with his toes warmed a little by the cylinders. The civilised motorcycle had taken a step forward.

This range of five models continued on until September 1977 when the 750 was replaced by a slightly larger model, the R80/7 of 800 cc. This was offered in two engine forms, the first pushing out 55 bhp on a 9·2:1 compression ratio and the second, built to satisfy the German insurance structure, with 50 bhp from a milder 8·0:1. It also had a different rear axle ratio and ran on regular petrol.

Otherwise it was the range as before in 1978 but with minor changes. The gearchange pedal was made remote so it could pivot in line with the top of the footrest and was connected to the box by a linkage with ball joints. The instruments were made easier to read and the rev-counter became electronic, while the addition of a cover between the handlebar grips made for a neater appearance and provided a mounting for any extra switches the rider might wish to add. Less popular was an audible signal that the indicators were in use. This used the same switches as the starter to cut it out in neutral or when the clutch was in so, as the rider moved off, so this bleeper would go into action. It clashed with pedestrian crossings with their audible tones for the blind, and was an embarrassment in a quiet village as it was much louder than the exhaust. Also unpopular was the provision of a first aid kit in a compartment at the front of the seat, for this

took away padding from where it was wanted and had an adverse effect on rider comfort.

The two sports one-litre models continued with their front twin discs, while the rest of the range including the new 800 had one. At the rear the RS was fitted with a new disc brake on the left with fixed Brembo caliper and both wheels were cast alloy as standard. Within a month or two the rear disc and alloy wheels were also fitted to the S, while the twin front discs were on the standard 100 and the R80/7. Finally the RS was made available in a metallic gold finish as well as the original silver.

During 1978 there were some changes in the range with the 600 cc model being dropped in the autumn. This left something of a gap at the lower end of the scale and one that had not gone unnoticed since the demise of the R50/5. To fill this space BMW came up with a pair of new machines, the R45 and R65, of 450 and 650 cc, and to all intents identical in appearance.

The smaller model was built in two forms, one with lower compression ratio and smaller carburettors to suit the German market and their insurance, the other for export. Both sizes of machine shared many common parts with the larger models but styled to disguise this, and both shared a crankshaft with 61·5 mm stroke. The small engine had a bore of 70 mm to give an actual capacity of 473 cc and the larger had one of 82 mm to give a precise 650 cc.

Compression ratio of both was 9·2:1 with the German model down at 8·2 and fitted with 26 mm Bing carburettors in place of the export version's 28 mm. The 650 used the 32 mm size. To help them fire-up in the morning an even more powerful starter motor of 0·7 hp was fitted. One useful change was to a self-contained points unit that keyed to the front of the camshaft and could be removed as one for attention. Behind the engines went the usual diaphragm clutch and five-speed gearbox as used on the other models.

The drive to the rear wheel was on the right as

Left **R65 engine, styled to look different from the larger models. The R45 looked identical to the R65**

Below **The smaller new twin, the R45 fitted with a pannier frame**

normal but contained a torsional damper in its length. This was of the spring-loaded cam-lobe type and drove rather lower axle ratios than usual to compensate for the smaller engines.

Both wheels were 18 in. diameter and both pressure die cast in light alloy. The front carried a single perforated 260 mm diameter disc with Brembo calipers with the option of a second disc being available. Due to the more compact nature of the new machines there was no longer room for the brake reservoir and master cylinder under the tank, for that space was taken up with the electrics. Thus the brake parts moved onto the right handlebar with the hand lever. At the rear BMW continued with their well tried drum brake with single leading shoe.

The new frame was more compact than the old and the suspension had less movement than on the larger models, although this still meant more than was usual for a medium capacity machine. As a result the wheelbase was shorter and very similar to that of the /5 series, while the width and seat height were both reduced. A 22-litre fuel tank with the screw-in filler cap was fitted.

The handlebars, instruments and controls were combined by a moulding that carried both speedometer and rev-counter with ignition key between them and covered the centre of the handlebars. The warning lights were set in the instruments and the control levers continued to be anodized matt black.

The minor electrical controls were completely revamped with the ignition key also turning the lights on. This left the right cluster free for the starter button and kill-switch and the left for horn, dip and flash, and turn indicators in a very neat and practical arrangement.

Right top **R100RT pannier showing off its fitted luggage bag. Too much weight is not advisable**

Right **Perhaps the ultimate BMW tourer, the R100RT with fuller fairing than the RS model. Clever paint work**

The finish of the frame continued black with colour for the tank, mudguards, seat tail and side panels in metallic red with gold lining, or metallic silver beige with olive lining. There was a long list of extras which still included the kickstart lever as well as the more expected crash bars, steering damper, twin horns, spotlights and extra instruments.

The new twins were not as fast as some of their competition but, with top speeds of 99 mph and 109 mph for the two sizes, were sufficient. In use the simple concept of the boxer engine once more came through as being what was needed for quick travel over a distance, especially if conditions were poor. Even the gearchange was good although a little slow at high engine speeds. The 450 was less satisfactory as it lacked the response of the larger engine as it had to pull along the same weight. It was still a very pleasant, if expensive, motorcycle.

An offshoot model appeared briefly late in 1978, the R100SRS. This was rumoured to have been a special export order that was either cancelled or only part delivered so the remaining machines became a regular model for a short time. It was based more on the S than the RS for it was fitted with the cockpit fairing carrying the clock and voltmeter. Added to it were crash bars, rear mudflap and hazard lights, while it was finished in white with tri-colour strips. It did not run on into 1979.

For that year BMW introduced a variant of their faired one-litre bike and introduced a few changes to all the larger models. The new one was the R100RT, similar to the RS but with a larger touring fairing with more upright and deeper screen, side pockets and air vents. It was built, in part, for the American market but had a few shortcomings for USA use. In that market the ability to carry was considered essential, for the touring American expected to bolt on plenty of extras and carry very full camping equipment. On the BMW they would go way over the gross weight limit so the lightness of the machine itself was no help. Neither was the continuing fitment of the first-aid kit at the front of the seat to the detriment of the rider's comfort. For European style riding with less luggage and a greater accent on handling it worked very well with easy 100 mph cruising without wind problems. Top speed was 118 mph and at that it was the slowest of the one-litre jobs.

Changes common to the larger models were on the lines of the smaller with the ignition unit, drive shaft torsional damper and switches all being adopted, along with a less popular move that kept the headlight on all the time the ignition was on. Not to everyone's choice and not helpful when cold starting with a tired battery. All the models now had the alloy wheels as standard.

The R100/7 model had a change of number to become the R100T and was fitted with the 65 bhp engine previously used by the S model. It was fitted with the S type seat now common to the larger models and with the tank in two colours as standard. It was also fitted as standard with voltmeter, clock, pannier frames (although the panniers were extra), high-rise bars, power socket, crash bars and a rear mudflap to go with its touring image. A comprehensive list of extras continued to be offered for all the models and on the smaller ones the extra front brake disc seemed to be the most popular as most machines were fitted with it.

For 1980 there were some additional colours, but otherwise the range continued as before until late in the year when the R80/7 was dropped and another 800 cc model appeared in its place. Normally such a change merely heralded some relatively minor modifications, but not on that occasion. The new model used the same basic engine but the machine and its field of use were totally foreign for a BMW

Right **Developed from the R90S, the R100CS gave the same performance and styling without the smoked finish. Back to Bing carburettors though, and Brembo calipers now fitted**

1978 model R65 in white. More sporting than the old R60 series

An R100RS far from home. Must be a test day, no pannier frames

production bike. For the first time they were to offer a trail machine but, true to their principles, it had a flat twin engine and shaft drive. It also had a disc front brake, a single rear suspension arm and unit, and a bright orange seat. It only weighed 167 kg, heavy for a trail bike but not for an 800 cc motorcycle.

Signs of the new model had been about for a while and one earlier report suggested a single cylinder machine with twin front discs coded R25. The same report did suggest that the twin would not make a serious off-road machine despite the good performance of the factory GS80 in the 1979 ISDT.

The new model was first reported to be an over-bored R65 as was the GS80 but in the event

Right **A convoy of dealers about to leave Munich on a fleet of R100RS models. On one such trip someone dropped one halfway across France but the BMW PRO coped**

proved to be the R80/7 engine on an 8·2:1 compression ratio and producing 50 bhp at 6500 rpm. It had a smaller and lighter clutch so could rev a little easier than the other engines and, while its twin Bings were as always, the exhaust system was a two-into-one device. The pipes were matt black and both ran under the cylinders along-side the sump with the right one crossing over under the gearbox to join the left and feed the gases into an odd black silencer that ran up the subframe bracing tube and along below the line of the seat to the end of the machine.

Internally the barrels dispensed with liners and had the bores coated with a hard alloy called

The R80G/S was the first trail BMW and proved to be one of their nicest road machines. Some development was done by Italy's Laverda factory

Galnikal and the oiling system was revised so that the mains were directly connected to the oil pump and not via the camshaft bearings. Ignition was electronic and the air cleaner simplified and lightened.

The gearbox was the stock unit but fitted with the kickstart pedal which was still an option for all the other models. It drove by shaft to the rear axle as usual but in a new form. The whole package went into a normal BMW frame with the long travel telescopic forks at the front. It was the

An R80G/S off-road and showing the true swinging arm rear suspension and black silencer

rear suspension that was unusual for the swinging arm was just that—a monolever as BMW put it—with a single massive member running back to the usual bevel box which bolted to it. Its movement was controlled by a single DeCarbon gas and oil damper and spring unit with five preload positions, and this lay forward at a considerable angle as was normal off-road practice and gave 170 mm of rear wheel movement.

The brakes were single disc at the front and drum rear, while both wheels had wire spokes and alloy rims. The front was 21 in. diameter and the rear 18 in., both being shod with studded trail tyres. The rear wheel was fitted straight to the bevel box hub with three bolts and so came off as a car one.

The fittings reflected the model character summed up by the type number R80 G/S, the letters standing for Gelanden/Strasse, the German for off-road/street. They also reflected a determination to get the weight down to a reasonable level for the rider seeking pleasure off the road rather than what was needed for serious enduro work. The mudguards were light and both were sprung, the forks had gaiters, the

Far from the Madding Crowd, **just BMW, rider and sheep enjoy the quiet**

headlamp was small and no rev-counter was fitted. A small console was built into the headlamp assembly and housed the speedometer, a cluster of warning lights, and the ignition switch.

The result was a machine that was really too big and heavy for off-road use but was able to cope well enough with occasional ventures into green lanes. On the road some thought it one of the best models BMW had ever produced. It was good for 104 mph and had extremely good acceleration, thanks to the light weight. The single disc was quite sufficient to stop it and the riding position quite acceptable despite the high bars which were finished black. They contrasted with the white of the tank, mudguards and side panels and also with the seat.

It went and it stopped and up to 90 mph it

The abbreviated console of the R80G/S without rev-counter. Works well and reduces weight

The R80ST which was derived from the trail model and retained many of its features

Right The sporting 650, the R65LS with mini-fairing and grab handles built into the seat. Much is new including the wheels

steered but then the big front wheel and the studded tyres began to have their effect and a gentle weave set in. Plenty of road riders immediately asked BMW when they would build a road version with standard bars and tyres but with no other concessions to road use that might put the weight up again. The one point left to help riders was the electric starter, kept standard on UK models, as the kickstarter was not the easiest to use due to its height above the ground. In full riding gear it needed a ladder to get the right purchase.

For 1981 the road range received a good number of changes, some already in use on the R80 G/S. In the engine all R100 models had the alloy barrels with Galnikal coating, while the small twins used the Nikasil process. A deeper sump was fitted to stop the oil moving about too much under heavy braking, and the R80 air filter and its black plastic housing was adopted throughout the range. All models were fitted with electronic ignition and an electronic voltage regulator, while the choke lever finally left the engine side and was neatly built into the left handlebar switch block as on the R65. The throttle cable linkage was modified to assist the carburettors to remain in balance more easily and a second exhaust balance pipe was added under the gearbox on the R100 series. The exhaust system for these models also had a new and even quieter silencer fitted.

For the small twins the R65 engine was fitted with a bigger inlet valve and the power was increased to 50 bhp at 7250 rpm, an increase that many thought would hone a fine edge on its performance. The latest clutch was fitted with reduced flywheel weight and with it came a smoother take-up and less effort needed at the hand lever. This, together with changes in the gearbox linkage, further improved the gear-change. Finally, in the box—when it was fitted—the kickstarter lever ratio was changed to make its use easier.

On the chassis side the rear fork of the small twins was lengthened by 10 mm, while that for the R100 series was stiffened. The bevel housing was changed to increase its oil capacity and rigidity, while reducing its weight. The disc brakes on all models became as on the small twins with fixed calipers, the master cylinder mounted on the right handlebar and a new pad material to improve wet weather braking was used. The fork style became common to that of the R65, while the battery was made a little easier to remove from the frame.

There were two changes to model numbers with the R100T becoming the R100, and the R100S the R100CS still with the cockpit fairing. A long list of options and extras was still offered including an oil cooler which was fitted as standard to the RS. One option that was fitted as standard to the RT was the Nivomat self-levelling rear suspension unit. This replaced the normal dampers and used gas, air, and oil to maintain a constant ride height regardless of the load carried. It did this by using the movement of the unit itself to pump it up until the factory set level was reached. The normal progressive damping was retained and the unit took from 300 to 1500 metres to stabilise depending on the road surface and the change in load.

A less technical but none the less equally welcome option was heated handlebar grips. At first thought of as a gimmick, all who tried them in cold weather found them very useful except in traffic when they were so good they had to be turned off. Apart from the motorcycles them-selves, BMW also produced a safety helmet to go with the rest of the range of clothing they listed. The helmet was a breakthrough for, although of the full-face type with hinged visor, the whole chin bar section could hinge up after the release catches on each side were operated. This enabled a spectacle wearer to don the helmet without removing his glasses—a very real boon. It was also possible to remove the chinguard to convert the helmet into an open-face off-road type.

BMW with single wheel trailer. Popular in Germany and Continental Europe but not allowed in the UK. Often seen in the Isle of Man during TT week

Late in 1981 a new variant of the 650 appeared, the R65LS. This was mainly a styling exercise with the machine being fitted with a combined cockpit fairing and spoiler whose lines were continued along the tank finish, seat and tail fairing. A new front mudguard was fitted and the wheels were totally fresh in concept and a composite casting. The star-shaped hub was pressure cast and the outer rim made of heat treated aluminium cast to the hub. The front was wider than on the L65 but it was still lighter, while the rear showed an even greater weight saving.

The styling exercise continued with the fitment of matt black exhaust pipes and silencers and dropped black bars. The seat had the grab handles built into the tail section so no separate rail was needed. The 50 bhp engine was fitted and twin discs at the front were standard. The finish was bright in either red with white wheels, or silver with the wheels to match.

In the middle of 1982 two new 800s appeared, one a road version of the R80G/S and the other a cut-price model based on the R100RT. The first was the R80ST which used the monolever rear end, the 800 engine and the sports cycle parts. The wheels retained the wire spokes, single front disc and rear drum of the off road machine but

Left **R80 for 1985 with alloy wheels and more conventional exhaust system than the R80ST**

Below **The first R80RT which used the 800 cc engine in the older one-litre chassis. For 1985 it changed to the monolever rear end**

Right **The new R65 for 1986 which used the R80 chassis**

The R100RS returned to the market for 1987 in this form, eleven years after its original launch, and still with the same format

with a 19in. front tyre and close fitting mudguard.

For road use the seat height and rear unit movement were reduced and the oil sump lowered. The result was a light, agile machine that was easy to ride fast and was at its best in the curves. Due to this the single front disc had to work hard and, as in the past, there were times when two could have been better even if heavier.

The R80RT was the second model and was a quick response to public demand for the 800 cc size machine. In effect it was simply a smaller sized version of the RT100RT with the same chassis and large touring fairing. It was also a good deal cheaper than the one litre machine, so it quickly found its place in the market. Top speed with the big fairing was around 95 mph and fuel consumption the wrong side of 50 mpg due to the effort needed to push it through the air. Thanks to a five gallon tank the tourist could still ride close to 200 miles before stopping, so this aspect was not too bad.

The whole range of flat twins ran on through 1983 and into 1984 before any real changes were made. The major BMW news during this period was the launch of a new concept of machine late in 1983, so it was inevitable that the twins should take a back seat.

During 1984 the new K series took over at the top end so the R100 models were all dropped. Otherwise the smaller flat twins ran on until late in the year when the R80RT was re-vamped into a new form. The major change was to the monolever rear end frame of the other 800 cc machines along with alterations to forks, cast wheels, exhaust system, final drive unit, stand and the electric system.

At the same time the R80ST became the R80 with similar changes and on both models the exhaust system comprised downpipes, an under gearbox collector box and low level silencer on each side. Only the G/S model continued with the single unit with high level outlet.

A little later at the beginning of 1985 a second version of the off road model appeared to celebrate the firm's continued success in the tough Paris-Dakar rally. It took this name and was fitted with an enormous 32 litre tank, a special single seat, luggage rack and crash bars.

During the year the smaller twins faded from the list but the R65 returned later in a revised form. This was a smaller displacement version of the R80 and used the same chassis. It proved to be an excellent introduction to the BMW way of motorcycling.

This gave a range of five flat twins in two capacities that ran on for 1986 and continued into 1987 with the return of an old friend, the R100RS. Demand for this well-faired, sporting model had been such that BMW found that they had to bring it back after a two-year absence.

It returned in modified form and as a limited edition of 1,000 units using the R80 chassis

Above **For 1988 the R100RT was also brought back to the range in response to popular demand for this long-range tourer**

Below **Only sold in Germany, the R65GS kept the older off-road chassis fitted with the derated R65 engine**

with monolever rear suspension. The sports fairing was fitted and the general measures to reduce noise and emissions included in the specification. Finish was in mother-of-pearl white-metallic with blue stripes, or in henna red with black stripes and black bottom fairing.

There were three new and one modified flat-twin model for BMW's 1988 programme, while the R65, R80, R80RT and R100RS continued, the last now a firm fixture in the range. The first of the new twins was, again, a return of an old friend, the R100RT. This had come and gone as the RS and, like that, returned in a familiar, but updated, guise with the monolever rear suspension and other improvements. In effect, the new model was the R80RT fitted with the engine from the R100RS. It came with an oil cooler, twin front discs, clock, voltmeter and panniers as standard in a Bermuda-blue finish with silver lines, and a black seat.

The other two new models were the R65GS and R100GS enduro machines, while the modified one was the R80GS, that was fitted with the new features of the larger enduro twin. The R65GS was only available in Germany and was effectively the old-style R80G/S fitted with the smaller engine, this being down-rated to 27 hp. This low power output was to suit the German licence restriction that limited a new rider to that level of power for the first two years on the road.

For the two larger models, the most noticeable new feature was a rear swinging arm patented as the BMW Paralever. It replaced the monolever arm, which continued on the road models, with a jointed design incorporating two universal joints and an extra link arm beneath the main one. The main arm continued to be pivoted from the frame with a universal joint in the drive shaft on the pivot axis. Just ahead of the bevel box a second pivot, plus universal joint, allowed the box to move, but under the control of the link arm which ran from a lug cast beneath the box to the main frame. Thus, the bevel box moved as if on an long arm with minimal angular charge relative to the main frame. The new design markedly reduced the effects of rapid acceleration or hard braking on the machine. No longer did the rear end squat down as the power was applied, very helpful to an enduro rider.

While the Paralever rear end was the major change, there were others for the R80GS and R100GS models. One was to Italian Marzocchi forks developed in conjunction with BMW from work done on the Paris-Dakar machines. Fork travel was increased a little from 200 to 225 mm, the fork diameter from 36 to 40 mm, the damper oil content increased, and its circulation improved for it to cope better with arduous off-road conditions. A bridge was fitted as standard to brace the fork legs and the 25 mm front wheel spindle was hollow to help reduce the unsprung weight. The front brake disc went up to 285 mm, as on the road models, and a larger Brembo calliper was fitted which resulted in lower operating forces and better feel for the rider.

Less obvious, but ingenious, were the new wheels that had tubeless tyres despite being laced with spokes between hub and rim. This was done by crossing the spokes in the normal way, as seen from the side, and also across the wheel itself. Thus, each spoke ran straight from one side of the hub to the other side of the rim at a point outside the tyre so the rim section

Above left **The R100GS introduced for 1988 had the Paralever drive shaft among its features**

Above **Smaller R80GS for 1988 that duplicated the many features of the larger model**

under the tyre was not pierced by spoke holes.

A further unusual feature of the wheels was that the spoke head was at the rim and the nipple in the hub. This allowed spokes to be replaced without removing the tyre and with the wheel still fitted to the machine. The effective wheel section was reduced where the spokes crossed, which helped the installation of the front brake calliper, and a nice fat rear tyre could be used.

All these new features went on to the R80GS, as well as the larger model, along with a lighter starter, bigger battery, larger presilencer, which reduced the noise level while improving the torque curve, and a 26-litre petrol tank. Unleaded fuel was acceptable thanks to a better valve seat material.

The R65 was dropped from the general range at the end of the year but continued to be

sold in Germany, fitted with the 27 hp engine. This was for the same licence restriction reason as the R65GS, that also continued and, thus, neither figured in the 1989 range. However, the other six flat twins ran on with one addition. This was the R100GS Paris-Dakar which was based on the standard version but incorporated many features from the works machines that had been proved in rallies. The model was not listed for the UK as it was fitted with a massive plastic petrol tank and regulations for that country insisted either on metal construction or EEC approval, a procedure hardly warranted for a specialised machine unlikely to sell in any great numbers.

All seven flat twins, plus the two for Germany only, continued for 1990 with minor changes. One common to all was a small increase in the width of the rear brake shoes and a revised shoe mounting. The GS models had a further option of sports suspension for the hardest of off-road riding, this being produced jointly by White Power and BMW. The

Above **Paris-Dakar version of the R100GS for 1990 with its various special fittings**

Right **New for 1991 for the enduro models was the Paris-Dakar type fairing, headlamp and protection bars as seen here on an R80GS**

option kit comprised a set of progressive-action front fork springs and an adjustable rear spring-strut designed for this type of work.

The special components that distinguished the Paris-Dakar model from the stock R100GS were available as a kit or single items, so any R80GS or R100GS could be updated as the owner wished. The kit comprised four main items, the first a 36-litre plastic fuel tank, either painted or left in primer for the owner or rider to decorate. Next came a fairing with a tubular support frame that provided fairing and headlamp protection and a lift point. There was further protection in the form of an aluminium sump guard, plastic cylinder guard and flared front mudguard. All served together to keep mud, rain and rocks away from both rider and

machine. For the R80GS the added guards made an oil cooler essential so the one fitted to the R100GS was used. Finally, there was a large solo seat that allowed a longer luggage rack to be fitted.

For 1991, the boxer range continued with the seven models, three for enduro use and four road models. The R65GS was dropped but the R65 ran on for Germany only. As was common for BMW, changes were really developments, rather than radical alterations, and honed the refined design still further.

The major improvement, which applied to all models, was called the secondary air system, or SAS. This introduced extra air into the exhaust port, near the valve, to burn off and reduce the more harmful emissions in the gases. The system involved an air filter hous-ing, two diaphragm valves and a coasting cut-off valve, while using the pressure pulses in the exhaust system to make it work.

Among the general improvements was a revised method of mounting single brake discs (already in use on the K-series), and a 27 hp version of the R80 engine offered as an option for any of the models. The option was to suit the German requirements and it was relatively easy to revert to the normal 50 hp when this was required.

The two enduro models had new looks for 1991 with the frame-mounted cockpit fairing from the Paris-Dakar model, its rectangular headlamp, and matching speedometer and rev-counter instruments. These went into a panel, the warning lights being fitted ahead of and between them, and the ignition key also central

Left **The R100R roadster in retro style introduced in 1992, combining Paralever suspension with a chrome-plated headlamp**

Below **First seen at the 1992 Cologne show, the R259 boxer engine for the future, true to its R32 origins**

and to the rear. The key was flanked by two rocker switches while the handlebar controls came from the K-series. The petrol tank cap was recessed flush to the tank top and fitted with a lock, the seat was improved, and a Bilstein unit controlled the rear wheel movement. The Paris-Dakar version of the R100GS continued as did the option kit for both standard enduro models. On the road, the R80, R80RT, R100RS and R100RT ran on, plus the R65 in Germany only. As in 1990, the R100RS was not available in all countries, while all models were little altered other than for the addition of SAS.

All seven boxer twins ran on for 1992 with the Marzocchi telescopic forks fitted to the road models as well as the GS ones. They were joined by the R100R roadster that adopted a classic style, harking back to the past, and was based on the R100GS. Thus, it used the same engine unit, the Paralever rear end, the wire wheels, but road tyres. To foster the image, the valve covers were rounded in form and came from the R68 while a chrome-plated headlamp shell was fitted.

By the 1990s, observers were predicting that the boxer was soon to go but had not reckoned on the innovation that came from BMW. Late in 1992 at the Cologne Show a new boxer engine was launched as project R259 and showed that the firm had every intention of keeping the engine type for the future. Conceived on a clean sheet of paper, the new twin was bigger and more powerful than any before, and introduced much that was fresh with four valves per cylinder, high camshafts, fuel injection and revised constructional details.

The engine dimensions were 99x70.5 mm, the capacity 1085 cc and the compression ratio 10.7:1. The crankcase was split on the vertical centre line, the two halves nearly identical, and the alloy barrels nickel silicon coated. Inside went a one-piece crankshaft turning in two main bearings while the connecting rods were sintered and forged in steel. This process gave close dimensional control and allowed for the big end boss to be fractured to provide the joint. This was a method already in use in BMW cars and gave a very precise fit.

Each alloy cylinder head had four valves and a single camshaft located alongside the valves with the sintered cams pressed into place. Each cam lifted a cup tappet, pushrod and forked rocker, that included the adjusters, to open two valves. The camshafts were chain driven from a jackshaft located beneath the crankshaft and itself chain driven at half engine speed. In this way the cams did not increase the engine width and their drive sprockets were of an acceptable size.

Cooling remained by air but around the

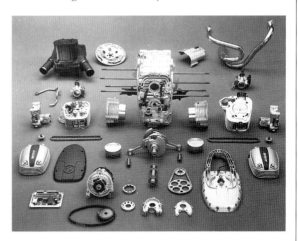

Above **Component parts of the new twin spread out for inspection**

exhaust valves it was assisted by oil. For this there was a second oil pump, designed more for volume than pressure, while the main one supplied the bearings, the system a wet-sump type. The jackshaft drove the oil pumps but the alternator went on top of the crankcase and was driven by belt from the crankshaft. It pumped out no less than 700 watts and an 1100-watt starter went on the left beside the gearbox. Ignition was electronic, a catalytic converter was fitted and the new engine produced 90 bhp at 7200 rpm. It drove the usual single-plate clutch and a five-speed gearbox based on that used by the K-series.

With this remarkable news, it was not surprising that the seven existing twins had little or no changes for 1993. However, there was one additional model, the R80R, essentially just as the R100R but with the smaller engine and no oil cooler which was not needed. The R80 was no longer listed for Germany and some other countries, and all the R80 models were available with the 27 bhp engine in Germany.

Spring 1993 brought the model that the new boxer engine was intended for, the R1100RS that introduced yet more innovation. Just as BMW had led the way to the telescopic front fork in 1935, so they now led away from it, but not to the hub-centre steering others had tried. The new BMW system retained the looks of telescopics but the sliding tubes were bridged and controlled by a frame-mounted wishbone and a single spring unit. Ball joints at the bridge and the top allowed a conventional degree of lock but the system was a true step forward and its design included a degree of mechanical anti-drive under braking loads. BMW called it Telelever.

For the rest, the R1100RS had a spine frame that comprised a cast aluminium front section

Left **The R80R of 1993, essentially a second retro model with the smaller engine**

Below **The R1100RS of 1993 that combined the new boxer engine with the Telelever front suspension – another radical advance**

The experimental C1 model that had extensive, built-in crash protection

and duplex steel tube rear. Rear suspension was by Paralever, the wheels cast-alloy in a three-spoke design and fitted with Brembo disc brakes and a second-generation ABS system. A sports fairing was fitted as standard and a full version listed as an option. An ergonomic package was another option available for the R1100RS and comprised an adjustable fairing windshield, adjustable handlebar rests, which had the option of being heated, and an adjustable seat.

In addition to this new machine that had gone into production, BMW also carried out a design study from which came their C1 experimental model. This sought to combine the advantages of car and motorcycle and had much the look of a scooter, but with the screen rails extended right over the rider. These linked the front and rear of the frame that was of a perimeter form, the whole combining into a roll cage with crumple zones to minimise accident injury.

Chain final drive for the first time since the Helios of the 1920s, the F650 Funduro of 1994.

A monster trail model, the R1100GS first seen in 1994 with the new engine and front suspension

The engine was to go under the rider's seat and might be a single-cylinder 250 cc four-stroke or 125 cc two-stroke, in either case driving automatic transmission. A catalytic converter, ABS, a safety harness and maybe an airbag were all on the agenda along with ample luggage capacity.

There were more new models for 1994, one of which truly broke from BMW tradition for, after 70 years, they listed a motorcycle having chain final drive. This was the F650 Funduro model that was built for them by Aprilia in Italy, and powered by a single-cylinder 652 cc Rotax engine from Austria, the machine having been first announced in June 1992.

The F650 was produced as a trail machine and its engine was conventional for 1994 in having four valves, twin overhead camshafts, water cooling and a balancer shaft. As standard it produced 48 bhp but was also available in a 34 bhp form to suit the new European restrictions, a rise from the German limit of 27 bhp. Twin 33 mm Mikuni carburettors supplied

the mixture, there were twin plugs to fire it and a catalytic converter was an option. Lubrication was dry sump and the oil was carried in the upper part of the tubular frame.

The five-speed gearbox was built in unit with the engine and the whole housed in a single-loop frame built up from pressings and tube, the lower part bolted to the upper. Telescopic front and pivoted fork rear suspension was used along with wire-spoke wheels and single disc brakes front and rear. Overall the F650 made a good package and BMW made sure that its style reflected the company and its image.

The second major new model for 1994 was the R1100GS, a massive trail model based on the new boxer engine and Telelever suspension. Its engine was detuned a little and the gearing lowered to suit the off-road use, while the suspension had more movement, the wheelbase was longer and the seat higher. The wheels were wire spoked and tyres more to suit the machine's purpose, while the ABS could be

This is the R100R Mystik of 1994 that had the twin front discs of the stock model plus a special finish

Larger of the new roadster models of 1995, the R1100R with Telelever, Paralever and new boxer engine

deactivated for that off-road use. Two front mudguards were fitted, one close to the wheel and the other frame mounted. The ergonomic package was fitted as standard as it was to the R1100RS.

Among the other models, all four R80 types were listed with 50 or 34 bhp engines to suit the Euro-regulation change, but the R80 itself was only listed for Germany and in the 34 bhp form. At the end of the year all four were dropped along with the R100GS. The R100R

gained dual front discs and SAS, and was joined by the R100R Mystik that had a distinctive finish and some detail changes to alter the style. The R100RT and Paris-Dakar R100GS rolled on as they were.

There were not too many changes for 1995, the R100RT, R100GS and both versions of the R100R all continuing as 'farewell models', a sign of their impending retirement after so many years of faithful service. The F650 in both forms, the R1100RS and R1100GS all continued with little or no change, but the twin range was amplified by two new models. These were the R1100R and R850R, essentially roadster versions of the GS. The larger used the GS engine, Telelever and Paralever suspension and the cast-alloy wheels from the RS although the wire-spoke ones were an option. No fairing was fitted. The smaller model copied the larger other than in its bore that was reduced to 87.5 mm so its capacity became 848 cc, and the gearing that was lowered.

Sure enough, the R100 models were dropped for 1996, but the R850R, R1100R and R1100GS all ran on without change while those for the R1100RS and F650 were minor, the single fitting a centre stand as standard. New was the R1100RT, created in BMW fashion as the grand tourer of the flat-twin range and therefore fitted with fairing and luggage system as standard. It used the engine from the R1100RS, had slightly lower gearing, Telelever and Paralever suspension, ABS as standard and came very fully equipped.

All the twins ran on for 1997 and only the R1100RS had any changes, these being to the front and rear suspension struts. The F650 single had a styling facelift with the fairing extended up but slimmer so that the turn signals went on stalks rather than in the fairing. It also had a taller windshield, lower seat, and an optional kit for taking the seat height down a further 50mm.

New was the F650ST, a road model based heavily on the existing Fundaro so that it used the same engine, transmission and chassis other than for a smaller front wheel and less rear suspension travel. Its style was much the same but it was strictly intended for road use. As with the off-road model, the new machine was available with the 34 bhp engine and the low seat kit.

So, the BMW Boxer ran on into the late nineties, ready for the millenium, while still closely following the lines laid down by Max Friz over 70 years earlier. The original ideals of fine engineering, good quality, and fitness for purpose were there as always, ready to continue into the twenty-first century.

Below **The R1100RT introduced for 1996 as the grand tourer of the flat-twin range, fully fitted out for travelling long distances.**

Top **This is the off-road F650 Fundaro in its 1997 form with revised fairing and taller screen**

Above **For 1995 the old boxer twins continued as 'farewell models'. This is the R100RT Classic that retained the lines first seen back in 1976 with the original R100RS.**

Right **New for 1997. The road F650ST was based on the Fundaro and used the same engine and gearbox unit.**

6 | Bright future

The near complete Krauser machine, the MKM 1000 with tubular space frame chassis and BMW engine. The four valve head is not fitted here

For a period of more than half a century BMW kept to a single concept for their motorcycles with the only variation from the flat twin being the single constructed along the same lines. Always the total machine clearly identified itself with the first model and for much of that time the company set itself the task of making machines for discerning riders who would, and could, pay well for them.

It is always a rather small, rather select market that has its good points and its pitfalls just as any other. Among the first is the tendency for the customer not to quibble about his bill. Just as with four wheels or even outside the transport field altogether this enables the work to be done completely and well, without skimping to save pennies. It operates in a world of 'if you ask the price you cannot afford it', of company vehicles and expense accounts, of leisure activities far removed from the daily grind—'pass down the bus' and 'please have the correct change'.

The pitfalls are many but the biggest and that fullest of past hopes and dreams is the reliability and service one. Owners who have paid a lot of money for their machines expect and demand either with voice or wallet that their spares are available, their machines serviced expertly and for them to be built to the highest standards. Even with money this is not easy.

The position with a motorcycle is further complicated by the normal view that it is a utility machine. This was to some extent retracted during the seventies when it was used more for recreation but always the idea of spending, for a motorcycle, the money that would buy a car was hard to swallow. The actual cost of the bike is also well within the budget of many more people so top bike owners could come from many more levels of society than top car buyers.

BMW had succeeded in this special corner of the market where many others had failed either because they were not in the right taste or because they were not good enough. In the case of the latter they tended to start off expensive, prove to be rather unreliable, become more expensive, lost sales and finally priced themselves out of the market.

BMW managed to keep just the right balance of selectivity while selling enough machines to keep the prices acceptable and a good network of dealers happy with enough turnover to win their support. This enabled them to be a force in every country they sold into that far outweighed their market penetration. They were ridden by many influential people, many in show business and many in the news, all of which helped to promote the company image and its product to others who could and would be inclined to purchase one.

BMW had remained the same in concept through many fashions. They had used unit construction from the start, been the first with telescopic front forks, quickly added plunger rear suspension, moved on to other means of controlling wheel movement, and finally introduced the integrated fairing. They had started in the day of the single with their shaft drive when the belt was only just giving way to the chain. Next had come the vertical twin with its inherent vibration, a feature not allowed in the boxer engine and near the end of the twin era there occurred the two-stroke triple with startling performance and enormous thirst. Then the time of the multi-cylinder four-stroke came, first with the four and finally the six. The four-strokes became more complex with twin overhead camshafts, four valve heads, sophisticated emission controls and finally turbo-charging. The two-strokes gained more ports, reed valves, power exhausts and narrow power bands. Frames became heavier to cope with the power and suspension systems more complex with rising rates, air forks and anti-dive features.

BMW continued with their flat twin, two valve, unit construction engine with shaft drive and year by year rumour had them building a new concept. As far back as 1958 the press had a racing four built and on test but nothing more was ever heard and the flat twin continued on its way.

Then in the early eighties came signs of changes. One indication was the announcement of a four valve head for the BMW by Krauser. These bolted straight onto the one litre engine and kept the standard carburettors and exhaust pipes. The valves were lifted by a pair of forked rockers moving on needle races and their clearance was set with a cam and locknut on each rocker finger. The compression ratio was increased to 9·8:1, the power to 80 bhp and the valve gear was safe to 7500 rpm.

This bolt-on accessory was only part of the Krauser activity for, in 1980, they unveiled the first of a limited production super sports machine based on the BMW engine unit. They kept all the running gear but fitted it into a space frame and covered it with a one-piece body moulding that ran from headstock to rear number plate and carried the seat. They also fitted a dolphin fairing and the result was a very special motorcycle that performed as well as it looked and offered everything that a rider wanted to experience from a machine. It was not for the town and it was very expensive but for the few who could afford such a luxury, and for those lucky to ride it, it wove a spell.

Around the same time BMW were taking a long hard look at their future on two wheels. At one point they even considered dropping bikes altogether but in the end decided not to. They

Complete Futuro with extensive fairing, disc wheels and turbocharged engine. Not too popular with riders

considered that, due to environmental problems, the rising costs of oil, and pressure from outside sources, the horsepower race was ending. The accent would be on lighter and more efficient machines making better use of their fuel, and in this their thoughts were reflected by trends in Japanese models.

In the late 1970s lightweight fours began to come from Japan, still very sophisticated but offering all the performance anyone really needed from smaller engines and with the bonus of improved handling and braking. The BMW concept of keeping their machines light and simple tied in with this and among their experiments was one done to explore the boxer engine layout as far ahead as possible.

To this end they commissioned an outside project team to take the flat twin engine and shaft drive, and to produce a design exercise model. The result was called the Futuro and was shown at the Cologne show in 1980.

The engine was the 800 cc unit from the trail bike with its power boosted by the addition of a turbo-charger mounted low down in front of the crankcase. Fuel injection and electronic ignition were used with micro-processor control and the gearbox was basically the stock unit.

The frame was part tubular and part mono-coque with the rear pivoted fork controlled by a single spring unit mounted above the gearbox. Front suspension was by telescopic forks. A very special body went onto this chassis and the rider sat in rather than on this. Instruments were all electronic and digital while a radio was fitted for those who preferred music to reading.

On the road the Futuro showed up the project team's lack of riding experience for although the machine had its good points comfort was not one of them. It was poor by any standards, much less the very high BMW ones. Also the machine was fitted with full disc wheels and just as in the 1920s, they proved to be unnerving in a cross-wind.

Back in the real world in September 1983 BMW launched their first new machine for 60 years and called it the K100. They had set them-

selves a very hard task with it for the model needed to be fast, needed to look different from the rest of the world and, most of all, needed to retain the BMW style of doing things. This had to reflect the way the machine performed on the road, its handling, gear changing, power curve, pulling abilities, braking and detail fitments rather than its appearance. It still had to stand out from the crowd.

BMW succeeded rather well in all this even if the end result did become known as the flying brick. For power they chose to use a four cylinder engine with twin overhead camshafts and to cope with silencing requirements they water

The BMW four cylinder in sports form with fairing. The K100RS at launch in 1983

Above **Basic K100 as launched late in 1983 – the first new BMW concept for 60 years.**

Right **The K75S with its cockpit fairing and engine spoiler as in 1991**

Above **Touring with a well-equipped K100RT fitted out with panniers and more extensive fairing**

cooled it. To meet the emission standards of the day and the future they elected to use an electronic fuel injection system linked to the ignition control circuits. The result was a broad spread of power so that only five speeds were necessary and as always the output went down a shaft on the right side of the machine.

The feature that made the BMW different from any other production motorcycle was the way the engine was installed. They first of all put the in-line cylinders along the frame in the traditional BMW way and then laid the engine on its side with the crankshaft on the right and the cylinder heads on the left. Ariel had done it before but only for a prototype.

This arrangement gave excellent access to most of the engine and a very low centre of gravity. Engine dimensions were 67 × 70 mm to give a 987 cc capacity and the compresison ratio was 10.2:1. The camshafts were chain driven from the front of the crankshaft and lifted the valves via bucket tappets. Two valves per cylinder were enough and a narrow valve angle kept the combustion chamber compact.

A gear at the rear of the crankshaft meshed with one on a jackshaft. This drove oil and coolant pumps which were housed together, the alternator and the clutch. The contra-rotation of these parts balanced out the effects of crankshaft acceleration and the clutch drove a three shaft, five speed gearbox of typical BMW layout.

Both ignition and fuel injection systems were of a sophisticated electronic design and incorporated safety devices to lower the power at 8600 rpm and shut the fuel supply off at 8750 rpm to prevent over-revving. With the peak output of 90 bhp coming at 8000 rpm this gave an ample margin. There was also a fuel cut-off in the system and an electric fuel pump in the petrol tank. The exhaust system was by four pipes into one square section silencer, all in stainless steel and the subject of some press criticism.

The frame was simple and tubular as the engine unit hung from it and contributed to the overall stiffness. Rear suspension was by monolever and front by telescopic forks. The wheels were cast in light alloy and of 18 inch front and 17 inch rear diameter while the brakes were

Brembo twin discs front and single rear.

The K100 fitments were interesting. Odd was the bar that hinged out to assist lifting the machine onto its rear stand. Odd also were the indicator controls with a button on each bar to turn the appropriate flashers on, plus an extra button to turn them off if the rider did not want to wait the pleasure of the auto system. Again odd was the provision of two warning lights, low and lower, for the fuel level.

For the rest it was the BMW package of good seat, large toolkit and nice detail fittings. The result on the road was rather unusual for the K100 behaved more in the way of a Japanese model as regards handling but BMW like in the way the power was delivered. A top speed over 130 mph was hardly slow but not really up to the hottest opposition of the day.

Really the machine performed very well for the more mature rider. It gave a more than ample performance without the need for constant gear changing and the only real complaints were the rather wooden feel of the brakes and a light front end at high speed.

At the launch of the basic model the K100RS was also shown. This had the sports fairing fitted which transformed the machine back into a total BMW. At speed the aerodymanics held the front end firmly on the ground to remove a small tendency of the K100 to feel uneasy about the front forks. The RS sat good and firm at 130 mph without qualm.

A little later the K100RT came along to complete the set and as was to be expected came with massive fairing and a vast range of useful accessories. With its big engine it could pull all this along with no problems and an acceptable fuel consumption so the result was a very good tourer for serious riders who also needed some performance for marching quickly through the boring bits.

With the three versions available the K100 showed it was a true BMW in the spirit of the twins and just two years after its launch BMW added the model everyone had expected of them. This was a 750 cc three cylinder version which used many parts of the K100. Two models were listed, these being the K75C with cockpit fairing and the K75S with the fuller sports type released early in 1986.

The engine layout was the same as the four except that balancer weights were added to the jackshaft to deal with the forces resulting from the three cylinders. Power output was 75 bhp at 8500 rpm and the engine had just about all the features of the K100 although the compression ratio was up to 11:1.

The gearbox was from the larger model as was most of the running gear and chassis. A Teutonic subtlety was the provision of a triangular section silencer in place of the square one of the K100. A change, which went onto the litre model, was a single warning light for low fuel level.

The gearing was lowered and a slightly smaller fuel tank fitted while the main difference between C and S versions was that the first had an 18 inch rear tyre and the second a fatter 17 inch as used on the K100. External variations came from the cockpit and sports fairings.

Just as with the K100, the K75 was slow in comparison with its contemporary Japanese competitors but its appeal was different. Its speed was more than adequate for any reasonable rider and it gave BMW comfort and sense of wellbeing. Allied to this was the knowledge that there would be no rapid changes and thus depreciation was retarded and the spares situation eased.

During 1986 the firm produced a limited edition as the K100RS Motorsport in celebration of the model being voted 'Motorcycle of the Year' for the third time by the readers of the influential *Motorrad* German magazine. The finish was in white, with a selection of stripes in orange, blue and purple, while the engine was black with some polished alloy edges.

Three more models joined the K range for 1987, these being the K75, K75S Special and the K100LT. It was also the year when BMW first showed their ABS anti-lock braking system although it was to be 1988 before they were able to offer it as a factory option for the K100 models.

The new triples used the existing engine, gearbox and chassis with the K75 becoming the base model without fairing and finished mainly in black with red or optional black seat. It was intended as an entry-level model while the K75S Special was effectively the sports model fitted with the optional engine spoiler as standard. Its finish was onyx-black with a pearl-beige seat or brilliant silver with black seat. In either case the wheels and engine block were black, this colour applying to the drive train and footrests of the silver finished model also. Engine fin edges were polished to make them stand out and add style.

The K100LT was a luxury version of the K100RT finished in Bahama bronze with a beige seat, claimed to be more comfortable than the standard item. As standard, the model was fitted with the self-levelling rear-suspension unit and, like the RT, with panniers and extended fairing. Additional LT items were a radio, bigger battery, hazard warning flashers, luggage rack, top box, soft rubber handlebar grips and the better seat. Both tourers had a further option in the form of an exhaust spoiler that fitted to the side of the fairing to direct hot air from the engine when riding on a hot day.

Another K100RS special edition appeared during the year as *Motorrad* readers had again selected the model as 'Motorcycle of the Year'. This time, the finish was in Avus black with a grey seat, polished fins, pinstriped edges and red 'RS Style' lettering on the fairing. As standard, the model had the sports-tuned suspension with reduced front fork travel and a modified rear-suspension unit. Pirelli radial ply tyres were fitted to both wheels.

Right **Entry level K-series model was this K75, first introduced for 1987**

Below right **The luxury K100LT was first seen for 1987 and included a radio amongst its added features**

For 1988 the major news was the introduction of ABS as a factory option for the K100 models. The UK price for this was £595 and it was May before it was available, but only as original equipment. It offered motorcycle riders, as long as they bought BMW, the same safe and predictable braking that had been available to car owners for a decade - and they could only overturn, not fall over!

BMW stressed that ABS was an aid that enabled a rider to brake as hard as any combination of tyre, road surface and weather conditions would allow - when this was needed.

It could not improve on the conditions, but gave the rider the best chance of braking hard in an emergency without locking up one or both wheels. In no way was it intended to encourage riders to leave their braking later, as if on a race track, it was to be used as a safety measure.

The system used both electronics and hydraulics with a 100-tooth impulse generator on each wheel and a pressure modulator in each brake line. A sensor scanned the teeth of each generator and was linked to a control unit that determined if either wheel was about to lock. If it was, the unit sent a signal to the modulator to reduce brake pressure until the risk of locking had passed, after which the pressure was allowed to rise again.

This cycle was repeated seven times a second down to 2.5 mph with the resultant system pulses kept to a minimum at the controls. Thus, there was minimal indication that the system was in operation, just sufficient to warn the

rider. The electronics went in the tail section of the machine and there were tell-tale lights to monitor the check sequence that the system ran through when the ignition was turned on.

If there was a system problem then lights would flash until the rider pressed a button that left a warning light on. This served to remind that the brakes were without ABS although still functioning in the normal manner. It was not a cheap option but was reliable and well worth having on any motorcycle.

The K series range for 1988 was as for the previous year except that the K75S Special was dropped. The other K75 models ran on with the K75S taking over the role of the Special by fitting the engine spoiler as standard along with the sports suspension. The normal suspension, as used by the K75 and K75C, was available as an option while the model had slightly wider handlebars.

For the basic K75 there was the choice of the two seat heights, these being 760 or 800mm (29.9 or 31.5in.), and detail changes in the finish. The same seat option was offered for the K100, which received a face-lift thanks to a number of changes. The radiator and headlight covers went, there were detail finish alterations and a high handlebar was fitted along with the 21-litre petrol tank from the K75.

The three faired versions of the K100 ran on as the RS, RT and LT, this last with a higher windshield with the option of side flaps. A lower screen was fitted in the model's home country, where riders had to be able to see over it, and this was an option elsewhere. With the new screen came an additional instrument panel carrying fuel and water temperature gauges, map light and cigar lighter. An optional top box was listed, able to carry more lug-

Left The K100RS for 1988, in this case fitted with the optional ABS system

Opposite **Rear wheel gear and sensor that signalled potential wheel locking to the control unit of the ABS system**

gage, as well as serving as a backrest for the passenger.

The RT and RS continued much as before and the K100RS Special Edition finish was in pearl white and Bermuda blue metallic with white cast-alloy wheels with blue pin striping, and a blue dualseat and engine spoiler. The screen was tinted black and the engine and drive train also black with the letters 'BMW' in white. The model came with the sports suspension, Pirelli radial ply tyres and an automatic side stand, the ABS being an option. Only 60 of the model came to the UK, half with ABS.

The K75C and K100RT were dropped at the end of the year but the other six models continued for 1989, to be joined by a new, and rather special model, the K1.

The K1 broke fresh ground for BMW for it was a supersports model that came in a bright Marrakesh red or deep Lagoon blue colour with yellow wheels, Paralever and seat edges. The model type, K1, was outlined very large on each side of the fairing and it added up to styling that seemed exuberant for BMW, although not so by some Japanese machine standards.

In fact, the finish was only the tail end of the K1 story for the fairing was new, the engine further developed with four valves and an 11.0:1 compression ratio, and the riding position aimed more at speed than touring. The engine remained essentially K100 with fuel injection but a single electronic management system was adopted, based on that of the firm's cars. An expansion box was added under the gearbox and a round-section silencer fitted. Fifth ratio in the gearbox and the overall gearing

were both raised while the rear-wheel drive used the Paralever rear arm.

The frame was a strengthened version of that of the K100 with Marzocchi forks and twin 305 mm discs at the front, these perforated spirally to reduce weight and braked by four-piston Brembo callipers. The discs floating in roller-shaped supports using a technique developed for motor racing. The Paralever was controlled by a new gas-pressure spring unit with more travel than usual, four pre-load positions, a progressive action for the spring, and damping effect related to strut travel. The rear brake was a single 285 mm disc. The ABS system was an option for the K1 from the start, but was modified to suit the specification of the model. Cast-alloy, three-spoke wheels were used, fitted with tubeless radial tyres.

Over this sound engineering went a new bodywork system that was the result of extensive wind tunnel testing and enclosed the engine. Turn signals were built in along with the instrument panel while vents allowed air in and out. Knee pads were part of the air flow system and were also to act as a safety measure to reduce the effects of a head on collision. The tank blended into the single seat, while the removal of a rear hump enabled a passenger to come along. However, luggage space was limited, especially two up. A welcome improvement was a long overdue central lock for both ignition and steering, common enough on other makes. It added up to a striking machine aimed at the sports rider.

For 1990 the K75 and K75S were offered with the ABS option and by August that year all K models sold in the UK had ABS fitted as standard. The basic K100 was equipped with a digital clock as standard but was no longer available in all markets, one where it was no longer listed being the UK. Both the K100LT and K1

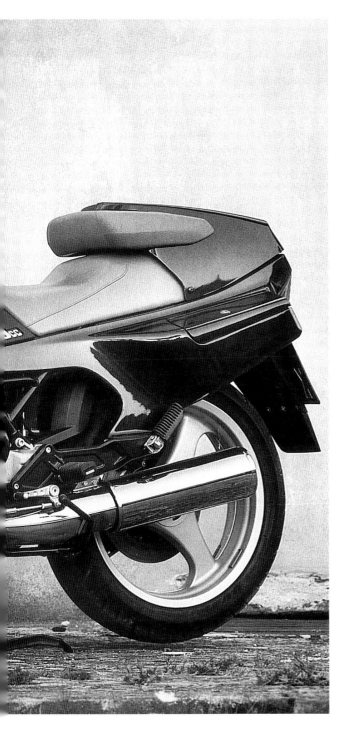

continued as they were but the two versions of the K100RS were combined into a revised form.

The 1990 version of the K100RS took many of the features of the K1, including the four-valve engine, raised fifth-gear ratio, front forks, front brake and Paralever rear suspension. Also used were the three-spoke, light-alloy wheels, stiffer frame and most of the suspension geometry. Variations concerned the rear suspension unit, taken from the K75S, and the overall gearing. Features that remained were the fairing, complete with its adjustable spoiler at the top edge of the windscreen, and the long list of options. As for the K1, the RS model had a central lock but its finish was slightly more restrained with a choice of pearl-silver metallic or yucca-green metallic, both having silver wheels and black engine cover and drive train.

One new model joined the range, but was only available for the USA and Spanish markets at first. The machine was the K75RT that was essentially the K100RT fitted with the smaller three-cylinder engine, lowered gearing and the fairing from the K100LT.

The K-series range stayed with seven models for 1991 but lost one and gained one. Out went the basic K100 and in came a Limited edition of the K100LT while the K75RT became generally available. The major news was a move to control exhaust emissions in an increasingly environmentally-aware world. BMW had been at the forefront, just as with ABS, with an announcement of a catalytic converter at the 1988 Cologne show. Two years later, again at Cologne, they showed the hardware and from May 1991 the K1 and K100RS were fitted with a fully-controlled three-way

Left For 1989, BMW introduced the K1 with revised four-valve engine and radical styling in a stark contrast to their more usual image

catalytic converter. For the other models, with less advanced electronic systems, a simplified converter was to be offered, one that gave good results but not to the same level of emission reduction.

In other respects the K-series ran on, the three-cylinder models the K75, K75S and K75RT. The four-cylinder range had two models with four-valve engines, these being the K1 and K100RS, plus two versions of the K100LT with two-valve engine. The standard one was as before with its extensive range of fitments but the Limited Edition version had a special paint finish, an engine spoiler and radial tyres, plus the sports suspension pack previously used by the older two-valve K100RS. This provided a fork stabiliser and modified front forks.

Although the three K75 models continued for 1992 along with the K1 and K100RS, the two K100LT machines were replaced by the K1100LT. This took the engine size out to 1093 cc by enlarging the bore to 70.5 mm, and adopted the four valves of the sports models. The rear suspension was changed to the Paralever system and a neat detail alteration was electric adjustment for the windscreen angle. The catalytic converter became an option for the K75 models while the K100RS had some changes but was dropped at the end of the year.

For 1993 the K75RT had the electric screen adjustment added to its option list and the

Top left **The K75RT appeared for 1990, for the USA and Spain only, reaching the general market the next year. It used the fairing from the K100LT**

Right **The K100RS for 1990, the year the model adopted the K1 engine. The engine spoiler was an option**

Above **Limited Edition version of the K100LT as offered for 1991, a fine way to go touring**

other four models all continued as they were. New was the K1100RS, replacing the smaller model, and fitted with the 1093 cc four-valve engine, a revised fairing, improved suspension and various detail changes. At the year end the K1 was dropped, its job now done.

In 1994 BMW were concentrating on their new boxer engine models and the Funduro so that changes for the K-series were few. One that applied to all was the fitment of the 700-watt alternator developed for the boxer, but otherwise there were no changes for the three K75 models and only detail ones for the two

K1100s except for the adoption of the second-generation ABS. The larger fours were joined by a Special Edition K1100LT that had a revised finish, altered seat and a radio with cassette player and remote control.

It was much the same for 1995 when there were no changes for the K75 machines, and only detail alterations to the K1100 models. At the end of the year it was announced that production of the triples would cease but that final editions of the K75 and K75RT would be offered during the first half of 1996. There were no changes for the K1100RS but it was joined by a special model having a two-tone finish. For the K1100LT ABS became a standard fitment while the Special Edition version had a revised finish.

The K1200RS replaced the 1100 for 1997, its engine having the stroke extended to 75 mm and the capacity to 1171 cc. It retained the features of the existing fours but was built in two forms, one producing 130 bhp, the other restricted to 98 bhp to suit the regulations in some countries. In either case the compression ratio was 11.5:1 and the engine management system further advanced. With it came the facility to plug in a diagnostic test set that monitored and advised on all the electrical systems of the machine.

For the first time BMW fitted a six-speed gearbox to a production model, the ratios closer and the housing smaller than for the five speeds. However, the general layout of the mechanics remained as before. A new frame was designed for the K1200RS and was die-cast in aluminium so that it fitted round the engine unit. Unlike the earlier K-models, it was not bolted directly to the engine but supported this through rubber mountings to isolate

Below **For 1992 this K1100LT replaced the two versions of the earlier model and used the four-valve engine**

In 1993 the K1100RS took over from the smaller model and appeared in this sleek form

any remaining engine vibrations. The frame comprised four sections welded together and was supported by Telelever front and Paralever rear suspension systems, the first time the new front fork had appeared on a K-model.

New, five-spoke, cast-aluminium wheels were used and the brakes were twin discs with four-piston callipers at the front and a single disc at the rear. ABS was fitted as standard as was a fairing carrying a tandem headlight of the latest design and a full set of instruments.

The K1100LT continued in two forms, the basic one no longer fitted with the panniers and top box as standard, these being options. In place of the Special Edition, the second type was the K1100LT Highline and this came fully fitted out as expected of a luxury tourer with its own special finish.

In all, the flying brick was a success and set to run on into the next century.

Left **The Special Edition of the 1994 K1100LT that came more than fully equipped for the long-distance traveller**

Below **The K1100LT Highline became the fully-fitted, de luxe version of the model for 1997, equipped to carry two in BMW comfort over any distance**

Below **The K1200RS of 1997 had a new frame, six speeds and the Telelever front suspension, being available in this stylish two-tone finish as well as single colours**

Bottom **Special model K1100RS for 1996 with two-tone finish to enhance its line and style**

7 | Solo racer

After their early racing experiences in the twenties, BMW returned in force to the scene in the late thirties with a supercharged version of their 500 cc competition flat twin. This had the Zoller eccentric vane blower attached to the front of the crankshaft which gave a much neater installation than the earlier Rootes type above the gearbox driven by the magneto. The Zoller was supplied by a single carburettor mounted to the right and fitted with an air filter. In time it was also cowled by a small alloy fairing to stabilise the air flow round it. The mixture was fed from the underside of the blower into pipes which ran under the barrels to the inlet ports and their length helped to cool the charge and damp out its impulses. It also acted as a reservoir which made throttle control a little more tricky.

The remainder of the engine followed the classic Rennsport layout with shaft drive to the two camshafts in each cylinder head, each of which controlled one valve via a rocker. The cams lay between the valves and, due to the cylinder offset, the drive shaft coupled to the exhaust camshaft on one side and the inlet on the other.

The low centre of gravity was a distinct aid to handling, while the two cylinders plus the blower gave enough power to deal with the Norton and Velocette singles and the Husqvarna and Guzzi vee twins despite the frontal area handicap of the flat twin engine. With it went the new tubular frame with telescopic front forks and, for 1936,

Famous scene showing Georg Meier about to start his
winning ride in the 1939 Senior TT

the plunger rear suspension. Otto Ley began to win races for BMW.

He continued to win a lot more races in 1937, along with Karl Gall, and that year one of the works bikes was allowed out with a non-German aboard, nearly unthinkable at that time but what Mercedez-Benz could do with their cars and Dick Seaman so BMW did with Jock West. West certainly played his part well to finish 6th in the TT and followed this up with a win in Ulster.

For 1938 BMW recruited Georg Meier from cross-country riding and he had an incredible season with four classic wins to become European champion. Gall and West backed him up with the latter coming 5th in the TT and again winning in Ulster. Meier went out of the TT with a silly fault. His BMW was warmed up on soft plugs

and one of the hard ones cross-threaded when they were changed. He made a token start and pulled off after a short distance.

In 1939 it was different and Meier and West dominated practice and the race to finish first and second. Thus Meier became the first foreigner to win a Senior TT, and BMW only the third foreign make, the others being Indian in 1911 and Guzzi in 1935. The works machines were the lightest in the event at an average of 306 lb or 139 kg, and one interesting detail was the use of coupled brakes controlled by the foot pedal via a small balance beam.

Meier won several races that year but then found himself up against the faster blown Gilera four and came off twice trying to keep up with it. In the end the European Championship went to the Italian machine and its rider.

Lovely period shot taken on the tarmac half of the Daytona road and beach circuit in the 1950s. Al Knapp rides the BMW, Don Kissinger the Triumph

After the war reports on the German motorcycle industry written in 1945/46 came up with some interesting sidelights. The power of the works blown 500 was given as 55 to 60 bhp at 7000 rpm as absorbed through the gears, so presumably measured at the output shaft or even the rear wheel. A less believable note in a US report suggests that road racing continued on the Continent into 1941 to keep Germany in the limelight, but this seems rather unlikely in view of their other pressing problems that year.

When the conflict was over, Germany was barred from international racing so continued to allow the use of blowers after the general ban at the end of 1946. Thus BMW were able to wheel out their 500 twin, and Georg Meier continued to win races on it, from 1947 to 1950 when Germany was re-admitted to the FIM. During this period he was well supported by Ludwig Kraus and in 1950 had several hard races against the blown NSU twin which ran mainly on methanol and whose 92 bhp made up for its considerable

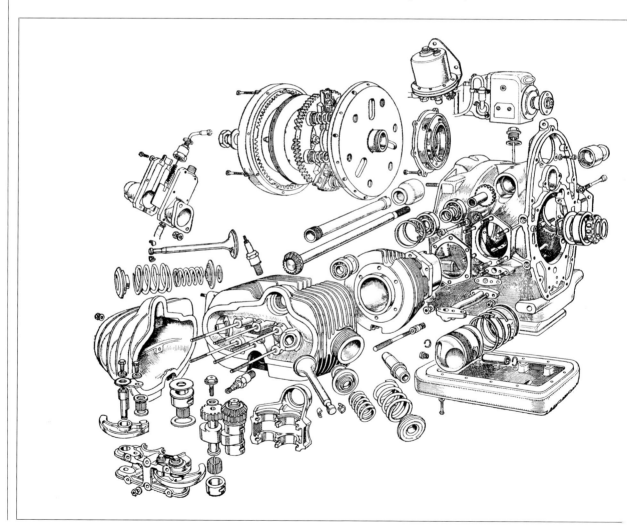

weight and dubious handling.

For 1951 the blower had to be discarded but BMW already had a machine running on carburettors and giving useful power. They also had a new works rider, Walter Zeller, who had been involved in the development of the naturally aspirated motor. Meier and Zeller began 1951 with this unit but by May had a new one with engine dimensions of 66 × 72 mm and running to 8300 rpm. The new BMW was shorter and more compact than the blown job and still

used the teles and plungers as before. Various cylinder heads were tried and a further rider, Hans Meier brother of Georg, was seen out on the machines.

It was good but not really good enough as the four cylinder Italian models were faster, while the agile English and Italian singles handled better and between them the flat twin could not make much impression. In minor events it continued to do well and Meier and Zeller came first and second in Germany's first postwar international

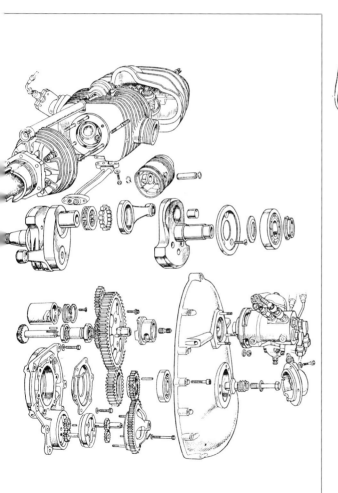

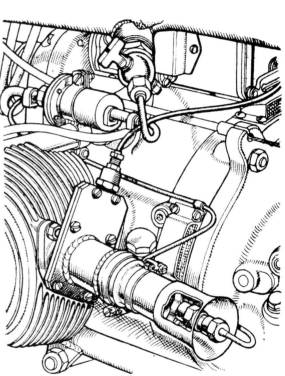

Above **Fuel injection on the BMW ridden by Zeller in the 1953 Senior TT fitted with guillotine slide**

Left **Line drawing of the Rennsport BMW showing the drive to the twin camshafts in each cylinder head**

Above **J. L. Lewis sharing the 4th placed BMW in the 1958 Thruxton production race**

Left **The great Fergus Anderson at Floreffe in 1956, the race in which he sadly crashed and was killed**

race at Schotten in 1951. Zeller also won at Hockenheim, but in the German Grand Prix at Solitude the best the BMW team could do was 5th and 6th behind the factory Nortons. In the middle of 1952 at Schotten a new version of the racing BMW was used by Meier fitted with swinging fork rear suspension. This was built by modifying a plunger frame and the drive shaft was run down the right fork leg. Meier won the event with Zeller second and later that year a more elegant frame appeared still fitted with the telescopic front forks. For 1953 the Earles type were fitted and a single entry made in the TT for

Right **Geoff Duke's BMW being fettled at Silverstone**

Below **Dickie Dale at Brands Hatch on his BMW in 1959**

Zeller. His machine followed the familiar lines of the flat twin with the magneto sitting on top of the crankcase and twin racing carburettors. Work was also in hand on a fuel injection system that year with the injectors fitted into the air intake mouths to direct the fuel straight at the guillotine throttle slides and inlet valves. Zeller's machine carried two spare plugs and a spanner in spring-clips and rather curiously the rear mudguard was unsprung like the front and followed the rear wheel contour closely.

In the event Zeller went out after a fall but was later seen in the German with a fuel system that injected directly into the combustion chamber. Late in the season one model was tried at Monza with a reversed cylinder head, an easy enough job to arrange on the BMW engine.

In 1954 Zeller continued with his works BMW, although still out of the hunt in classic events. He came second in the German to Duke's Gilera at the Nürburgring and in this event John Surtees also had a BMW ride but was forced to retire due to a misfire. Zeller's machine gained a hydraulic steering damper in 1955, then considered a novelty but to become standard wear for most road racers in time.

1956 brought the appearance of Fergus Anderson on the flat twin but sadly he was killed at Floreffe when he fell late in the race. Zeller still rode and that year was at the TT once again. He no longer used the discontinued Rennsport type engine, but a short stroke version with dimensions of 70×64 mm that produced 58 bhp at 9000 rpm when fitted with carburettors and 61 bhp with fuel injection. It had five speeds in the gearbox and in this form the drive shaft ran alongside the fork leg, not inside it. Unfortunately a last minute decision to dispense with the streamlining left Zeller overgeared for the race but, despite this handicap that kept him from using top gear at all, he finished 4th.

The days of the solo racing BMW were drawing to a close as far as the 500 cc Grand Prix were concerned but it still did well in other events in the hands of Ernst Hiller, Ludwig Kraus and Zeller.

In 1958 came the news that Geoff Duke was to ride the works BMW which was based on the machine previously used by Zeller, and that Dickie Dale would also be on a Munich twin. Duke won at Hockenheim after a close race but retired in the TT where Dale was 10th. Dale managed some places in the classics during the year but Duke had less success and several retirements.

While spares for the works engines were hard to come by there were always the older models and, when Walter Zeller retired at the end of the year, BMW presented him with a road bike. But what a bike, for it was a 600 cc blower engine unit fitted into an RS frame with telescopic forks and road equipment. A good tool for the autobahns.

Dale continued to campaign the BMW in 1959 but with limited success and the next year the Japanese, Fumio Ito, rode them with style but with little hope of success in the classics. The day of the solo flat twin had never really come in the postwar era as, without the blower, it was hard put to overcome the handicap of its frontal area and two cylinders always had to bow to four in terms of power output.

BMW had also been active in the early days of production racing where their reliability paid off. As far back as 1948 they took the first three places in the 24-hour Bol d'Or and managed a third the following year. In 1954 the first 24-hour race to be run in Australia took place and a 600 BMW was first with a 250 cc single taking that class.

In England in 1958 an R69 was fourth in the Thruxton 500 and impressed all who watched with its silence, its consistency, and its few regular stops for fuel and rider change. The same year a 500 was second in the Bol d'Or and in 1959 repeated this, while that year saw the R69 successful both in the Thruxton 500 and the 24-hour Barcelona race. It was the first time a large capacity machine had won the latter event for the twisty Montjuich circuit suited small capacity models. The same machine nearly pulled it off again in 1960 but time spent in the pits allowed a 175 cc Ducati ahead. The make did however take the first two places in the Bol d'Or, the last to be run for some years.

In 1961 the BMW was very successful in these long distance events, the machine being an R69S entered by the MLG Motorcycles concern. In May they won the Silverstone 1000 km event, and at the beginning of July the 24-hour Barcelona one after trailing a Ducati for much of

Right **Daniels at the end of the Barcelona 24 hour race won by a BMW in 1961**

the day. Later that month the BMW also won the Thruxton 500. After that MLG sold off the machines and quit racing for a while.

Ten years later, in 1971, the solo BMW was once more running in production events and was in the Isle of Man for the TT. The result was a reasonable fourth for Butenuth riding a R75/5, with another one 9th. In 1972 Helmut Dähne repeated the fourth position and again the following year. That same year a BMW was third in the Bol d'Or. In 1974 BMWs were second and third in the TT but a solo win continued to elude

Bruce Daniels at Silverstone on the MLG Motorcycles entered BMW. The first of three victories that summer

them, while from then on they had to give best to the multi-cylinder challenge from the Japanese factories with, first, Kawasaki and then Honda dominating the endurance events.

In the USA the flat twin continued in a very special form prepared by Butler and Smith. They ran in the 1972 Daytona meeting using a worked-on R75/5 engine fitted with one or two factory items and installed in a lightweight frame. Krober ignition and 35 mm Mikuni carburettors

gave it 9000 rpm and a reasonable amount of power. A 250 mm Fontana drum brake was fitted in the telescopic front forks, while an American disc went on the rear hub.

By 1975 the Butler and Smith BMW, ridden by Reg Pridmore, was a very special machine indeed. All road-going features were removed and engine castings extensively modified to reduce weight and improve performance. The most cunning move was to reduce the width of

Left **The Butler & Smith BMW in close company with another rider in a US event. Well modified and well ridden**

Below **Helmut Dähne in an Isle of Man TT in his familiar leathers**

the engine by fitting special short titanium connecting rods to the standard stroke crankshaft. With Venolia pistons in the barrels these could be shortened but the rocker covers were still ground away on the corners so had welded-on flats to fill up the holes.

The head only contained two valves but had twin plugs and gave a high compression ratio of the order of 12:1. Valve rockers were totally stock and not even polished as this tended to

lead to them breaking. Otherwise it was a case of careful development along known lines and detail improvements. One such filled the flywheel with holes and reduced its weight by more than half.

Behind the engine went a five-speed box, at first a Kaiser unit based on BMW parts, but later a BMW box was used. The drive was still by shaft to a lightened bevel box unit. The engine unit was installed in a Rob North frame with duplex tubes throughout and a massive swinging fork still pivoted on taper rollers with the drive shaft in the right leg.

Suspension was by Ceriani fork legs in Betor clamps at the front and Girlings with two-way damping at the rear. The wheels were cast magnesium from Morris and fitted with tubeless tyres, Dunlop on the front and Goodyear at the back. Brakes were well drilled discs, twin at the front and single rear.

The result was a 165 mph motorcycle weighing 310 lb and handling well enough to worry its competition. It was not really able to worry the two-stroke opposition but did go well enough to keep them on their toes. The sound was better as well.

The experience also gave them an edge for the 1976 Daytona Superbike Production event when they built three bikes using many of the same techniques, headed the qualification table, and finished one—two. Again they used short rods to narrow the engine down and many special engine parts in the R90/S unit. A monoshock rear end was used controlled by a special racing car Koni unit set up for the bike. The machine also won that year at Laguna Seca and Riverside.

They were less successful in 1977 when they managed a 4th at Daytona, but the rest of the field had speeded up and stayed ahead for 1978 as well when the BMW was third.

In 1981 BMWs reappeared at Daytona in a new style event—the Battle of the Twins—

Left **Endurance racing BMW in a 1974 24 hour race. Machine comfort aided rider to maintain pace throughout the whole event**

Up she comes—a BMW running on butane gas takes off in 1981. Note single arm rear suspension

evolved to bring back a host of big capacity four strokes, all with distinctive sounds. All manner of machines turned up from Harleys to Vincents, including both production type BMWs, one of which came fifth, and a Rennsport which came seventh. It was an event to savour as it brought back evocative noises to the race track and once again the hard rasp of a flat twin in full flight was to be heard.

The heyday of the solo racing BMW was really in the period before the war, as after it their success was always tempered with a condition. 1938 was perhaps their best year, with the TT win of 1939 the icing on the cake. Afterwards classic success eluded them, although the machines were always competitive at international level.

On three wheels the story was a little different.

8 | Sidecar racer

The solo racing flat twin may have had its problems in terms of frontal area and torque reaction but, once on three wheels, the tables were turned. The twin gave more power than a single and the flat twin had several advantages over a vertical one. The cylinders spread themselves well out into the airstream either side of the front wheel and did not hide behind it. It was nearly in perfect balance so vibration was no problem at all which reduced the likelihood of parts fracturing during a race, and gave the driver and passenger an easier time of it. Thus they could concentrate on directing the outfit, not in trying to restore life to tired muscles. The low engine unit allowed the driver to adopt a lower riding stance and, thanks to the outline silhouette of the unit, he could easily move about over it when cornering. The carburettors or air intakes could be long and set for downdraught without them getting in the way however much the rider hung out.

Possibly the greatest asset the BMW engine unit had for racing was the combination of its unit construction and shaft drive. The engine and transmission were a single unit, with a five-speed gearbox, that could be installed in the frame as one item. From it the shaft drove to the rear axle bevel box at the rear wheel and the shaft joints accommodated any misalignment as the chassis heaved and bucked over the circuit bumps with no more trouble than it dealt with the normal wheel movement on its suspension.

Hillebrand followed by Schneider and the pack in the
opening stages of the 1957 Sidecar TT which he won

Walter Schneider with passenger Hans Strauss pursued by Fritz Scheidegger who won this one after Walter retired with water in the ignition

In contrast, most of the single and vertical twin engines opposing it had a separate four-speed gearbox driven by chain and a second chain to the rear wheel. No matter how well they were braced when constructed, the engine, the gearbox and the rear wheel would all move about under racing stresses. Unlike the BMW, the constituent parts moved against each other instead of combining to improve their structural rigidity. The result was chains running out of line and a lack of oneness to ease the driver's task.

The BMW formula for sidecar success was to

work for a long, long time.

Up to 1950 the machines were confined to German events and supercharged to compete against the blown NSUs, as in the solo class. In 1951 they reappeared on the international scene, and at Hockenheim their leading driver, Wiggerl Kraus, hung on to the great Eric Oliver for the whole race to pull out and cross the line with him. That year Oliver was to win his third world title and was seldom headed except by Frigerio or Milani with their much faster Gilera fours. To keep Oliver in sight was a real achievement, and later that year at Solitude Oliver led Kraus until his sidecar wheel spindle broke to force him out and let the BMW through.

On the classic scene in 1952 there was little

Florian Camathias at Scarborough in 1963

sign of the Munich twin, while in 1953 Oliver took his fourth and final title, but Kraus, Noll and Hillebrand all managed places in classic events using twins with carburettors or fuel injection.

In 1954 the tide changed in favour of BMW for although Oliver won the first three classics, Wilhelm Noll won the other three after Oliver had suffered a broken arm. Noll also had a second place and this was sufficient to take the title. That year there were no makers' titles awarded but had there been it would have taken a time tie-breaker to settle the issue as BMW and Norton both had three wins, two seconds and a third from the six races. In the TT BMW were only bettered by Oliver who was followed home by Hillebrand, Noll and Schneider.

The next year a BMW won all six classics with three going to Willy Faust, two to Noll and the TT to Walter Schneider. Faust took the title and BMW their first maker's award and one they were to hold on to for two decades.

Noll won his second title in 1956 using an outfit with very extensive streamlining in which he sat with his legs enclosed. That year Hillebrand won the TT and BMW lost a sidecar race for the first time since Belgium in 1954. Milani brought the Gilera out at Monza and Pip Harris also got his Norton home ahead of the twins. The Italian repeated the exercise in 1957 with Cyril Smith second on a Norton, but BMWs won the other events and Fritz Hillebrand won both the TT and the title. Sadly he was killed late in the same year whilst racing at Bilbao.

1958 only had four classic chair races and Walter Schneider won three of them including the TT with Camathias second on each occasion. Only in Holland did the Swiss manage to reverse the position. It was Schneider again in 1959 with Camathias pressing harder so that each driver took two races, while Scheidegger won from the champion in France. 1960 saw BMW again make a clean sweep of the classics with Helmut Fath winning four of the five races and finishing behind Pip Harris in Holland.

Fath and Camathias suffered serious accidents in 1961 which allowed Max Deubel to win his first title, but any suggestion that he and his partner Emil Horner did not have the necessary ability was completely thrown out as they went on to take the title three more times to create a unique four in a row score. They won three of the races in 1962 and for once a BMW missed a classic win when Chris Vincent and his BSA took the TT after both Deubel and Camathias retired when well ahead. A BMW driven by Kolle was second. 1963 was harder with Camathias winning two races, as did Deubel, and Scheidegger the fifth but Deubel had two seconds to keep the title.

In 1964 it was again a hard struggle between these three and again Deubel managed it with two wins and seconds, including a TT win. Finally in 1965 he had to give best to Fritz Scheidegger who won four of the seven classic rounds and was second in the other three, one of them being the TT which Deubel took. In 1966 it was Scheidegger all the way for he won all five events with Deubel second in four of them.

Sadly Scheidegger was killed early in 1967 due to a crash at Mallory Park and Camathias had already gone at the end of 1965 in an accident at Brands Hatch. Deubel retired after the 1966 season so the field was open for new men and

the one who emerged from the pack was Klaus Enders who won five of the 1967 classics to take the title. He was partnered by Ralf Engelhardt in the sidecar, and in addition to their title they were second in the TT behind Siegfried Schauzu.

In 1968 BMW ran into a problem in sidecar classics in the shape of Helmut Fath. He had crashed badly in 1961 and, while convalescing, had planned to build his own 4 to beat the flat twin. By 1968 it was ready and he won three of the six classics which was enough to gain him the title, although BMW retained the makers one. Once again Schauzu won the TT in which Fath was fourth.

For 1969 it was Fath versus Enders who managed to retake the title with four wins including the TT, while Fath crashed in the penultimate Finland round which put paid to his chances. He then retired and his absence enabled Enders, passengered by Wolfgang Kallaugh for most of the year, to take his third title by winning five of the eight events, including another TT. He followed Fath into retirement while Fath's four cylinder machine was driven by Horst Owesle in 1971. He only won three classics but these taken with two seconds gave him the title, although once again the makers one stayed with BMW. Schauzu had also won three events including yet another TT but had only one second.

Owesle retired after his title win and Enders came back to sidecars after a rather unsuccessful season with BMW cars. He won four classics to take his fourth title, but in the Isle of Man it was Schauzu who once more won. Enders won a fifth title in 1973 and totally dominated the classic scene with seven wins from seven starts, including the TT in which Schauzu had to be content with second place.

Times were changing, however. The first glimpse of a possible threat to their sidecar supremacy appeared in 1972 in the shape of the two-stroke engined Konig ridden by Steinhausen. In 1971 BMW machines and the Fath

Above **Fath at Ballaugh Bridge on the TT circuit with all three wheels off the ground**

Right **Max Deubel also airborne at Ballaugh in the 1965 Sidecar TT which he won**

Above **Sid Schauzu going well**

Right **Klaus Enders and Ralf Engelhardt in the Belgian in 1972. Still a 'pudding basin', in 1972!**

URS 4 had dominated the leader boards of the classics. In 1972 they continued this but a Konig was third in the TT. By 1973 the Konig presence was very obvious for, despite the total success of Enders and his BMW, he was followed home by a Konig in second or third place in every event.

It was very apparent that the four cylinder Konig was faster, especially at the start of a race when it was not too hot, and that Enders' skill was playing a large part in keeping him ahead. The basic construction of the BMW flat twin with the shaft drive was still playing its part and, as always, the lack of vibration helped to keep the driver and passenger fresh in a long race. The two-strokes lacked some of these advantages for the Konig and other similar units were designed for boat use with a constant supply of fresh cooling water. They did not take kindly to a recirculating system with radiator and pump. In

addition, the engine was not laid out for a motorcycle transmission, a separate gearbox had to be used and chain drive employed. Although lighter, the outfits now had more power so all the familiar problems of alignment and rigidity arose despite the use of clamps and guides to hold the separate parts in place.

If 1973 had shown the coming menace, 1974 confirmed its arrival. For the first time a two-stroke won a classic sidecar race, the German held at Nürburgring, and Enders and Schauzu chased it home. At first it seemed as if it would be Schauzu's year for he won two of the early races, but gradually Enders sorted out his engine and by a narrow margin Klaus won his sixth world title. It was a fitting climax to two decades of success for the Munich flat twin as no other firm had ever dominated a racing class for so long or so decisively.

From 1954 to 1974 BMW won 19 of the 21 riders' championships, and 20 out of 20 makers' titles. The missing year is 1954 when only riders' ones were awarded, while in 1974 the picture was confused by Enders' machine being entered as a Busch–BMW. Konig gained the official title as the BMW wins were split between the usual name and the special Enders model. The flat twin in fact won enough events for the title.

During the period from the German GP at Solitude in 1954, won by Noll, to the Dutch TT at Assen in 1974 where Enders was first, there were 128 classic sidecar races. BMW won 112 of them and their main opposition came from the home-built Fath 4 which accounted for 10 of their defeats. Others were due to the Gilera 4 on three occasions and BSA, Honda and Konig, the Honda success being from a private owner at Finland in 1973 where none of the usual riders took part. It is a truly astonishing record only ever matched by the MV company in the solo field. BMW also won the sidecar TT from 1955 to 1974 with the sole exception of 1962, and also the over 500 cc sidecar TT from 1969 to 1975.

Their success was due to the superb driving of

Enders and Deubel, who between them accounted for more than half of the titles, with Schneider, Noll and Scheidegger taking a pair each and Faust, Hillebrand and Fath the remaining one each. In the TT Schauzu, nicknamed 'Sideways Sid' or 'Siggy', won four 500 cc races and another five of those open to larger machines. His sidecar TT record is unmatched. Enders, Deubel, Schneider and Hillebrand also left their mark on the TT courses, both Mountain and Clypse, but most of all it was the flat twin with its sense of fitness for the task that was so successful.

After Holland in 1974 there were to be no more four-stroke successes and in 1975 none finished in the first three in any classic race. Sidecar racing had become two-stroke dominated and was to rapidly move from boat engines to the 500 cc Yamaha engine with its unit construction, although the final drive remained by chain, and then into far more sophisticated chassis developments. At the same time noise regulations clamped down on the

Enders in the wet showing the low build reached by the early 1970s, very different to earlier days or to the shapes to come

Fritz Scheidegger in full flight running without a fairing

four-strokes and racing lost one of its great attractions.

Right at the end Mike Krauser, better known for motorcycle luggage equipment, prepared a very special BMW flat twin sidecar with the engine fitted with four valve cylinder heads and twin, belt driven, overhead camshafts. Unlike the normal engines, the intakes were splayed out from the top of the heads, not behind, and the exhausts emerged underneath. Fuel injection was used into the inlet tracts but, despite all the work and the immaculate preparation, the outfit was unable to beat both the two-strokes and the noise regulations.

The BMW had been equally successful in international and national events for it made a very good racing sidecar outfit. They were not easy to come by for not many were made and both machines and spares were expensive. However, once set up they would run through a season with no great need for attention so were well suited to the turmoil of national road racing with events every weekend.

Their advantage was not just speed for it was possible to tune other engines to be as fast, it lay more in their stamina, lack of vibration, broad power band and total suitability for the job. The BMW engine was slow to warm up but did not tire in a long race, usually speeding up a little. The outfits normally had coupled hydraulic brakes on all three wheels so the hand lever was never used and this allowed the driver to concentrate on steering and throttle opening.

They won a lot of races and were always magnificent to watch and listen to. The engine note had a purposeful whine to it that made all that heard it confident that the flat twin was in command of the situation. And so it was for many, many years.

9 | Speed records

BMW stood at the pinnacle of record-breaking during the 1930s for it was during that period that they attempted and captured the absolute motorcycle speed record not once but many times. Just before they retired from their road racing activities of the 1920s they turned to this aspect of two-wheeled sport and began a decade of competition with English and Italian men and machines.

Their first attempt came in 1929 when the record stood to the credit of Herbert Le Vack at 207·33 km/h and had been set at Arpajon in France. His machine had been a 1-litre Brough Superior and the record was set after a day-long feast of speed during which the old record had been broken four times.

Against this BMW wheeled out a 750 cc flat twin fitted with a supercharger. Power output was 55 bhp and some attempt was made to fair the machine with a cowling over the tank which extended back to the rear wheel. The cylinders were still out in the airstream but, with Ernst Henne aboard wearing white leathers and helmet, the machine was fast enough over a stretch of new autobahn from Munich to Ingolstadt. His speed was 216·05 km/h and the date September 9. It was the first time a German machine had held the record.

Henne returned to the fray in 1930 just three weeks after Joe Wright had raised the record to 220·99 km/h. Once again he used his smaller blown engine against the 1000 cc vee twins

Right inset **Ernst Henne with the streamlined helmet he wore on more than one successful record attempt**

Below **The 1929 record attempt with the 750 cc supercharged machine. Cycle parts from the R32**

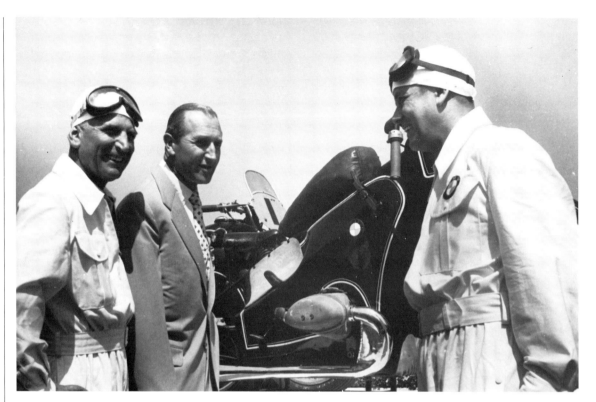

favoured by his rivals, and with the fairing on his machine extended to discs on the rear wheel he pushed the record on a little more to 221·52 km/h. Joe Wright replied with 242·59 km/h on the Carrigrohane Road in County Cork, Ireland, and although there was a controversy as to the make of machine used, the satisfying point to English riders was that the 150 mph barrier had been passed. It was also a sizeable increase and an unusual move for a rider weaned on the Brooklands tradition of moving on in small steps with a payment bonus for each one. It was not the done thing to add over 10 mph in one go.

Henne took the sidecar record at 190·83 km/h in 1931, the chair being a hemispherical section with totally exposed third wheel on the right of the machine. The attempt was made at Vienna along with a solo run of 238·3 km/h which was up on the previous year but not by enough. Both attempts were made with the blown 750 cc machine.

In 1932 he went to the Tat road in Hungary in November with the 750 carrying further fairings to enclose the cylinders. Even the bevel box

sported a tail fairing and, while the machine with its trailing link, leaf sprung, front suspension and rigid rear may have seemed odd, it represented the best knowledge of its time. It may have looked as if little advance had been made since the debut of the R32 in 1923 but Henne came back with the record at 244·399 km/h.

In 1933 he tried with a sidecar again but without success, although by then the sidecar wheel was driven as well as the rear one. The next year he went to Hungary again and moved the figure on to 246·059 km/h, while in 1935 he ran the 750 up and down the autobahn near Frankfurt to pass the 250 km/h mark at 256·046 km/h. The machine did not look much different from 1929 but the engine was developing about 100 bhp at 6000 rpm. The streamlining still consisted of panels added to the bike, while Henne sat on top of the device wearing a faired helmet and conical tail held to his body with straps.

Although the machine, used in 1935 to take the motorcycle speed record, looked crude and reminiscent of the 1920s, the one BMW wheeled

Left top **Ernst Henne centre with Wiggerl Kraus on the left and the 1929 record breaker at a post war function in Munich**

Left **The fully faired record breaker at the 1937 speed festival attended by Auto Union and DKW as well as BMW**

Right **For the record attempt Henne ran without the enclosure and wore his faired helmet. Here he is pushed off on a run up the autobahn**

Jack Frost setting Australian records in 1957 at Coonabarabran, NSW, at 149.06 mph

out for Henne to use in 1936 was anything but. It still had a flat twin engine and shaft drive but that aside it was all new. Scorning the use of capacity the engine was the racing 500 cc unit with Zoller blower mounted on the front of the crankshaft. It developed 105–108 bhp at 8000 rpm and was fitted into a tubular duplex frame with telescopic front forks and plunger rear suspension. The whole machine and its rider were totally enclosed by a fairing with detachable top that allowed Henne to get in and out. The fairing left the lower third of the wheels exposed, had air inlets near the nose for cooling, had an exhaust pipe poking out of each side and sported a pair of stabilising wheels for use at standstill speeds. The record moved on to 272·006 km/h and was set on the Frankfurt to Munich autobahn.

It was the turn of other riders. First Eric Fernihough, well known at Brooklands, went to Gyon in Hungary and managed to raise the solo record to 273·244 km/h with his unfaired Brough Superior. He had been unimpressed by the BMW fairing but his subsequent run with a sidecar bolted to the Brough showed that faith and a powerful engine had really become too much of a handful. He did take the sidecar record but onlookers were surprised that he survived.

Later, in 1937, another fully faired machine attempted the solo record. This was a supercharged 500 cc Gilera ridden by Piero Taruffi and he had already taken the one hour record at 194·8 km/h by riding up and down the autostrada between Brescia and Bergamo. This was despite the need to slow nearly to a standstill for

Right **The record breaker on show with the shell half cutaway to show the racing blown 500 cc engine, and record breaking type chassis**

the turn at the end of each leg. In October he went out on the Brescia road to capture the world record for Italy for the first time with a speed of 274·181 km/h.

Sure enough Henne and BMW reacted and late in November they were back on the autobahn between Munich and Frankfurt. The machine used had run earlier in the month at a speed festival attended by the incredible Auto Union cars as well as motorcycles and BMW were not alone with their full fairing. DKW had both semi- and fully-enclosed models, the latter proving to be highly unstable. Henne rode his twin in its 1936 form with cover over his head, tail fin and air brakes at each side near the tail. While it looked stable, Henne reported it to be a handful to the extent that it could not be held on full throttle.

For the attempt the tail fin was removed and the machine run without the top enclosure, while Henne reverted to his white faired helmet. The attempt was completely successful with the record being taken at 279·503 km/h, a figure it was to remain at until 1951, when a blown 500 cc NSU in a shell not unlike that of the BMW finally moved the figures on. In his attempt Henne also broke the one mile, five miles and five kilometres records, all flying start ones, and his figures stood in the 500 cc, 750 cc and 1000 cc classes. Quite a good day's work for a small flat twin with shaft drive.

After the war, on the 10th anniversary of his successful attempt, 28 November 1947 Henne tried once more on the autobahn near Frankfurt but on that occasion failed to better his old speed. Despite that he could look back on an unrivalled record for to all intents he had held the title of fastest man on two wheels since 1929. On each occasion his figure was broken he soon retrieved it, and in all had broken the absolute figure seven times.

It was May 1954 before BMW returned to the record breaking business and for their target they chose to attack the eight and nine hour records

using a 500 cc RS fitted with a full front fairing which gave the rider plenty of protection but left the cylinders out in the air-stream. The riders were Georg and Hans Meier with Walter Zeller, and one of the features of the machine's fairing was a cover that concealed a pair of lights for use at night. The attempt was made at Montlhèry and was successful at 166·64 km/h, or just over 100 mph.

Late in October the team was back at Montlhèry with rider Wilhelm Noll and a very fully faired sidecar outfit. The motorcycle was completely encased except for the wheels at the road, rocker box covers, megaphone ends and the rider's back. Three struts connected the machine to a sidecar wheel, also fully covered with a slot for the tyre, and the main fairing had two air inlets placed high up just under the screen. Fitted with the 500 cc engine it broke the 10 km record at 212·7 km/h in 500, 750 and 1200 cc classes.

In March 1955 Noll, Schneider and Hillebrand were all at Montlhèry for more long distance attempts using the 500 cc RS with the front fairing and covered headlights. On this occasion it was coupled to a sidecar platform and took 57 world time and distance records in the three sidecar classes. These ranged in distance from 500 km to 2000 miles, and in time from 5 to 24 hours. The average speed for this last time, one of the most difficult of all records, was 144·3 km/h. The firm thus held nearly all the recognised long distance and period records for sidecars, the ones not held being for even longer runs.

Later in the year in October BMW made further record attempts using the Munich to Ingolstadt autobahn and two machines, one solo and one sidecar. The latter was driven by Noll and further developed from that used the year before, being fitted with a cover over the driver

Right **Willy Noll and the fully streamlined record breaker used in 1955**

The 24 hour record breaker at Montlhèry with Bruce Daniels aboard

and a tail fin. The cover could be jettisoned in an emergency by a quick-release device. Noll was successful and his unsupercharged 500 cc engine took him to a new world sidecar record at 280·22 km/h. He also broke a number of other short distance records.

The solo was a standard road racing machine fitted with full front fairing only and was ridden by Zeller over the 10 km and 10 mile distances to break the record at 241·8 km/h over the longer one. It retained its carburettors unlike Noll's engine which had fuel injection.

The factory were not alone in their search for records as in 1957 in New South Wales, Australia, Jack Forrest raised that country's 500, 750 and 1000 cc records to 149·06 mph using a BMW with very full front and rear streamlining. The rider sat in rather than on the machine with just the rocker covers in the breeze.

Late in the next year Florian Camathias established a new sidecar record for the 100 km distance at 194·8 km/h. He did this at Monza using the banked circuit on which he rode clockwise so the sidecar wheel, without any enclosure, lay above him on the left on the

banking. It was a trip not without hazards for the front tyre finished with much of the tread torn from it by displaced streamlining and was down to the canvas. Previously Camathias had over-turned while practicing due to a front tyre failure.

In 1961 a BMW was used to set new 12 and 24 hour world records at Montlhery. The machine was an R69S fitted with an engine prepared by Munich. Machine preparation was by the London based MLG concern who fitted a Peel fairing, a racing seat, padding on the tank and other suitable items. How special the engine was internally, was indicated by the need to drain the sump and pour in hot oil before a cold start. No air cleaner was fitted and the carburettors had the usual BMW style long bellmouths, while on the exhaust side open pipes with megaphones were used. Ridden by a team of four it took the 12-hour record at 109·34 mph, the 24-hours at 109·24 mph and covered a total distance of 2621·77 miles.

Twelve years later a pair of BMWs ran further and longer in a Maudes Trophy attempt in the Isle of Man. This trophy was awarded by the ACU to a company whose standard machines took part in a certified test of sufficient merit. It had been presented in 1923 by George Pettyt of Maudes Motor Mart, Exeter, and was originally called the Pettyt Cup. The aim of the award was to encourage demonstrations of the reliability and economy of motorcycling.

From the start some arduous tests were devised with the Cup going to Nortons for four years from 1923, until Ariel won it. It moved about to various English firms in the thirties, finishing with Triumph where it stayed for a good while. In 1952 BSA took it from them but again there was a long period before an attempt by Honda in 1962 took the trophy abroad for the first time. Since then no test had been considered severe enough to warrant its award.

BMW chose to run a pair of R75/5 machines round the TT circuit for a week. The bikes were taken at random from the production line in

Start of the Maudes attempt on the TT circuit. Geoff Duke and the Mayor of Douglas flag off the first two riders and the rain began to fall

Berlin, sealed and run in for about 1000 miles, not perhaps long enough to really free the engines off. The models were the final, long wheelbase version of the /5 series fitted with the factory options of halogen headlight, crash bars, twin spotlights and large tank. A team of 14 riders were involved.

The attempt began at mid-day on a May Thursday and, as the Mayor of Douglas and Geoff Duke flagged the machines away, the rain began to fall. It was to continue, backed by mist and winds, for most of the next seven days. One rider crashed and one had an accident in torrential rain, which meant cycle part replacement but

the engines ran on and on. One clutch had to be changed due to the hard riding and lots of tyres, but otherwise the machines just kept going. By the end of the week one bike had covered 217 laps and the other 225 to make an aggregate total of 16,685·55 miles at an average speed of 49·57 mph. Due to the weather the test turned out to be very tough on both men and machines, and the riders, mainly journalists, all thought that the BMWs, and they, had done enough to warrant the trophy being awarded.

The ACU thought so too and later in the year came the announcement that BMW had been awarded the Maudes Trophy.

It was a fitting finish to their record-breaking and came in their 50th year of building motorcycles.

10 | Six Days and sidecars

BMW were involved with the ISDT as far back as 1933 when the event was held in Wales due to the British victory the year before. It was a close run thing and during the third day German hopes received a set back when Ernst Henne lost a mark through having to stop to repair a puncture. They revived later that day when a British team rider dropped two marks, and at the end of the week it was the German men who had held onto their tiny advantage to take the Trophy. With Henne in the team was Josef Stelzer on the second solo with Mauermaeyer and Kraus on the sidecar. The machines were based on the flat twin in the pressed steel frame and fitted with trailing link forks.

The 1934 event was held in Germany and once again the trio of Henne, Stelzer and Kraus on their BMW twins were successful so Germany again organised the event in the following year. For that they entered the same three riders but on very new machines. These had the racing type overhead camshaft engine complete with its crankshaft driven Zoller supercharger on the front of the crankcase. The frames were fitted with the new telescopic front forks and the whole team were very keyed up to try to win the Trophy for the third consecutive year.

Kraus had the hardest time and turned his sidecar outfit over twice. The first time his passenger, P. Muller, was knocked out, dumped back in the chair and finished the few miles to the finish of the day's run unconscious. On the

second occasion he collected a broken rib but kept going to the end of the week. By then it looked as if the Trophy would go to Czechoslovakia as their team on Jawas had a slender lead over the Germans and just had to keep going through the speed test for victory. However, it was not to be—one of the engines dropped a valve and the Trophy remained with Germany.

They were unable to repeat their success the next year when Britain regained the Trophy, so in 1937 the ISDT was again held in Wales. The German team were as usual mounted on BMWs

ISDT BMW as used by Meier, Zeller and Roth in 1952 when the German team was second in the Vase

and comprised old hands Stelzer and Kraus with newcomer Georg Meier. At the end of the week they tied with the British team so the result depended on the speed test run at Donington Park. After a hard ride the Trophy went to Britain by the very narrow margin of 10 seconds.

In 1938 the Germans revised their tactics and machinery with the solo men on DKWs and only

179

Right **Hans Roth in the speed test in the 1958 ISDT**

Below **Typical ISDT picture captures the flavour of the event with a BMW rider following a competitor**

their sidecar entry, Kraus, on a BMW. It was a bad Trophy year for Germany as two DKWs went out on the second day and Kraus broke his sidecar axle. However, they did take home the Vase, the club team prize and the Hühnlein trophy introduced that year for military and similar type teams.

Despite the international tension of the following year the ISDT was run, and in Germany, after the ACU had declined it. It took place in August and at first all seemed well. Then in the middle of the week tension mounted and the British riders left to ride for home. The trial results were declared void.

After the war the ISDT got underway in 1947 in Czechoslovakia without even the British team, but in 1948 they were back in Italy at San Remo to capture both Trophy and Vase. They kept the Trophy for the next three years, by which time BMW were back with a maker's team and among the gold medal winners at the end of the week were Roth on a 250, Georg Meier and F. Larsson

on 500 cc solos, and Kraus with his sidecar. Zeller missed his gold by a single mark to win a silver with his 500 cc twin.

In 1952 the German Vase team comprised Meier, Roth and Zeller all on 600 cc BMWs and up to the last day it seemed they were certain to take the award. Then at the start of the speed tests Hans Roth was unable to get his engine to run and the resulting loss of marks dropped the team to second place. The trouble turned out to be a broken valve in the left cylinder, a poor reward for such a good week's riding.

1953 saw the BMW trio as half the German Trophy team but their team chances were wrecked on the third day when the universal coupling on Zeller's machine began to break up. This put him well behind time and forced him to retire. Meier and Roth both completed the course without loss of marks to win their golds and, while no longer a team member, Kraus again won a gold with his 600 cc sidecar.

The next year there were no team BMWs but in 1955 the West German Vase team used them with Roth collecting a gold, Kraus a silver, but Hans Meier being less fortunate. Among the silvers on a BMW was Sebastian Nachtmann, later to become an ISDT star. For a number of years there were no BMWs in the teams, but in 1958 Hans Meier, Roth and Hartner all won golds on their BMWs, and in 1960 Nachtmann was in the Vase team with his 600, often setting the pace in the special tests for bonus marks.

In 1961 West Germany won the Trophy once again and, as in earlier days, a BMW was in the team. It was Nachtmann who rode it and he was back in the team the following year but had the misfortune to crash on the first day and crack a bone in his right forearm. In pain, he managed to finish the day but although he started the following day his stiff arm forced him to retire.

Nachtmann rode his BMW several more times, including in 1965 as a West German team member, and was nearly always well up with bonus points but the day of the four-stroke was

really well past. Small two-strokes could be pushed, dragged or carried if necessary to get to the finish, but big heavy twins had to be ridden all the way.

The 600 used by Nachtmann retained the Earles forks up to 1962 and was an impressive and weighty machine. It was fitted with a saddle and adapted and modified for ISDT work but was still a real handful to handle on the rough, especially on downhill and muddy routes. On the road it was magnificent with the sidecar gearing fitted improving the acceleration while still allowing it to easily reach 90 mph. The trail tyres hardly encouraged anything faster. Externally there were changes to the mudguards to prevent mud clogging them, to the exhausts to tuck them out of the way, and to the wheel spindles which had tommy bars brazed into them. A hydraulic steering damper was fitted and its effect adjusted in the same manner as was to be used for the /6 road machines. In all it was a very good machine, although not as well suited to its task as some of the big British twins. It emphasised how good an ISDT rider Sebastian Nachtmann was.

In 1963 he was equipped with telescopic front forks a good few years before the road models became so fitted. In time this work led to the lightweight road prototype already mentioned, and the equivalent ISDT model used in 1965 by Manfred Sensburg had much in common with the forthcoming /5 series as far as the cycle parts were concerned. The leading axle front forks used then were very similar indeed to those used from 1969.

By 1967 the factory were no longer directly involved but that year assisted Klaus Becker in his preparations for the ISDT held that time in Poland. He still used the early type of swinging fork frame but fitted with the front wheel, forks and tank from a Yamaha. The engine unit, however, bore markings indicating that it had been in the BMW factory test shop.

In later years even larger BMWs came to be

Left **1969 BMW set up for off-road work**

Below **The Hughenden BMW motocross outfit built in 1974. The rear bevels could not take the strain of the jumps**

used, with Helbert Scheer riding a 750 in the 1970 event. He used it well enough to collect a silver at a time when few riders opted for a four-stroke at all, or for anything approaching that capacity. The 1971 event was held in the Isle of Man and two BMWs were there, one ridden to a gold by Herbert Schek and the other by Kurt Distler. He was most unlucky as, with a silver seemingly in his pocket, he crashed two miles from the finish of the last day's run and was unable to take part in the speed test.

1973 saw a factory team once more when the ISDT was held in the USA, and Schek, Kurt Tweesman and another rode their 750 cc BMWs in the Vase team. For a while they held an advantage and Schek put up the fastest times in the acceleration tests. Unfortunately all the BMW riders lost points due to the combination of the short travel of the kickstarter and the reluctance of the engine to fire in the cold mornings. This pulled them back and any hopes they had died with their team-mate's Maico gearbox.

Schek had been involved with the ISDT and BMW for a good few years and had built up a useful relationship with the factory as a good rider, a mechanic and a dealer with a large business. The result came to fruition in the middle 1970s as the GS80, the largest and fiercest ISDT BMW ever. The engine was based on the R80 unit but fitted with a very carefully balanced crankshaft with the R65 stroke of 61·5 mm. With a bore of 95 mm this gave the very over-square and not too wide engine of 872 cc that ran up to 6700 rpm to produce 57 bhp on a 9·5:1 compression ratio. It would pull smoothly from 1500 rpm and the tremendous torque allowed it to trickle round bends in third and surge away down road or track. Internally the engine was special with Nikasil coated alloy barrels, single compression

Hughenden then built a grass track sidecar which was smart and successful

The GS80 run by the works in the 1979 cross country championship. From this came the R80G/S. Note monoshock

ring pistons and R80 cylinder heads each fed by a 32 mm Bing.

The frame was duplex and light, while the rear fork was controlled by a single unit on the drive shaft side. This attached to a welded assembly of tubes on the fork leg so the damper unit could lay down and connect to the frame just beneath the seat nose. At the front went Maico forks modified by BMW and carrying a drum braked front wheel from the same source.

The resulting performance with an all up weight of 140 kg was very spectacular indeed, although it took a big man to turn the engine over to start it. Many thought it too big, too heavy and too fast for ISDT and enduro work but much of this was the thought of the power, weight and width across the rocker boxes. In

practice it worked well and only at very slow speeds was the weight apparent.

The GS80 ran well in the 1979 ISDT held in Germany with Fritz Witzel taking the over 750 cc award and was part of the team that won the Vase in the 1980 event run in France. From it sprang the R80 G/S designed for trail use and one of the nicest of road BMWs. It was also good enough to win the 1981 Paris to Dakar rally over 6000 gruelling miles. The winning machine was ridden by Hubert Auriol, and two other BMWs finished in 4th and 7th places in this tough event. All three had long-range fuel tanks and strength-

ened chassis to cope with a three-week event.

Auriol repeated his success in 1983 with Belgian Gaston Rahier taking the honours in 1984 and 1985 to give BMW four wins. In 1988 the qualities of the marque were further exhibited when Eddy Hau won the marathon class for private owners, one where replacement of nearly all essential components was prohibited in the tough rally.

On three wheels, Hughenden Motors, a BMW dealer in Newbury, used a modified R90/6 engine in a Wasp frame to create a side-car moto-cross outfit.

Unfortunately the rear bevels protested about their treatment in moto-cross over the jumps and this led to the development of a

Hubert Auriol on the BMW in the Paris to Dakar rally which he won in 1981

grass track outfit using a well modified R100RS engine. This had its compression ratio raised to 15:1 to suit the methanol fuel and big Dell'Orto carburettors. Twin plugs in each head, fired by individual coils and electronic ignition, provided the sparks. The shaft drive was modified to incorporate a shock absorber and some car parts were used in this area

All told it worked very well to show how successful the boxer engine could be in an environment far removed from its gentleman's sport tourer image.

11 | *Wehrmacht und Polizei*

Motorcycles are used by the authorities in most countries in both war and peace so it is not surprising to find military examples of the BMWs built in the late 1930s. For the army it is often necessary to push machines out in a very short timescale and in most cases these finish as being one of the simpler and sturdier models from the civilian lineup, shorn of all essentials and painted in a single drab colour.

In England this led to the use of the 16H Norton, M21 BSA, C and CO Enfield, and so on as they were available and easy to maintain in the field. An oddity was the Flying Flea Enfield, used by the airborne troops, and dropped by parachute in its own crate to give the men far more mobility in the crucial early hours of an attack. It was light enough to be manhandled over obstacles but all the other machines were rather heavy, low powered and used to carry messages and act as convoy escorts.

The *Wehrmacht*, likewise, had its similar machines, many of them BMW, fitted with panniers and headlamp mask but otherwise little changed except for colour. Some were used with a sidecar and in some cases this would be fitted with a machine gun. As well as the BMW they were also supplied with Zundapp flat twins and NSU singles but in 1944 the big four-strokes were

Engine unit of the R75. Note in-line kickstart, frame construction and small silencer box

replaced by two DKW two-strokes, one a 125 cc single and the other a 350 cc model.

As in other countries, some of the machines used were impounded civilian models, the BMWs being the R5, R51, R61, R66 and R71. However, the main military models were the single cylinder R4 and R35, plus the twin R12 and R75. Both the R4 and R12 were really obsolete but, as in other countries, the services were expected to manage with them.

During their campaigns the Germans examined captured Belgian FN and Sarolea models, also the French Gnome et Rhone, all of which had sidecar-wheel drive and were able to cope with the mud. This led to a deci-

sion that BMW and Zundapp should both build similar sidecar outfits and the result was a complex design in either case.

Later, it was realised that small, light and cheap machines were better and could cope with a myriad of duties. Until then the Germans had their flat twins with four-speed gearbox, auxiliary two-speed gearbox, reverse gear, sidecar wheel drive and lockable differential.

The R75 engine was an overhead valve version of the R71 unit and retained the 78 mm bore and stroke. Compression ratio was 5.6:1 and on this it produced a modest 26 bhp at 4000 rpm. It was supplied with mixture by a pair of 24 mm

Graetzin carburettors and they in turn were supplied with air either from a filter mounted on top of the gearbox or, for desert use, by a larger one on top of the tank. The tank top filter contained a removable element under a domed cover and its mounting incorporated a choke worked by a lever protuding from under the cover. Suitable hose connected filter and carburettors.

On the exhaust side the pipes ran forward and down to a transverse cylindrical box from which a pipe ran back along the length of the right side lower frame member to the rear bevel box. There it turned up to enter the centre of a high mounted tubular silencer with rear outlet and perforated heat shield. When the machine was in service in the cold of Russia the exhaust was passed through an exchanger to warm air which was ducted to large hand shields, to the rider's feet and into the chair. This went someway to combat the bitter cold.

Ignition was by magneto mounted on top of the crankcase and driven by the gear train at the front of the engine which also turned the camshaft. A pancake dynamo was bolted onto the crankshaft nose and protected by a round cover on top of which a smaller round cover hid its cut-out unit. The lubrication system was standard BMW with wet sump, gear pump and strainer on the pick-up in the oil.

Behind the engine went the usual single plate

Right **Wartime R75 fitted with machine gun and panniers to sidecar. Headlamp mask typifies the period**

Below **Several R75 machines in convoy during 1942. Second machine in line is an R12, the side valve, pressed steel frame model**

dry clutch and this drove the double gearbox with its four speeds and reverse plus high and low ratios. The main gearbox was controlled by a foot pedal on the left and this was supplemented by two levers on the right side of the tank and one under the saddle.

The inboard tank lever moved in a gate to select forward or reverse, the latter protected by a catch released by depressing a button in the lever knob. The second tank lever selected the high or low gear ratio and the one under the saddle locked the differential between rear and sidecar wheel drives for maximum traction under adverse conditions, although this made the steering much heavier.

A curious reversion occurred with the kick-starter for, although mounted on the left, it swung in a normal arc along the machine not transversely so represented a throwback to the earliest days of the company.

The differential and sidecar wheel drive were all built into the rear bevel box which was driven by the normal exposed shaft and universal joint. Within the box were 12 gears and the output for the sidecar wheel was positioned ahead of the rear wheel spindle. A substantial coupling allowed the sidecar drive to be detached along with the chair, while in use it connected to a shaft running across the chassis within a tube. At its outer end it carried a gear and this meshed with another mounted ahead of it on the sidecar wheel axle and by this means the required wheel lead was obtained.

All three wheels were large with very heavy gauge spokes to withstand the weight and cornering loads. They all carried 4.50×16 in. block pattern tread tyres and all interchanged while a spare was commonly carried on the sidecar tail panel. All three wheels had brakes about 250 mm in diameter with cable operation at the front and coupled hydraulics for the rear and sidecar. There was no external mechanism on the front backplate and the cable entered to connect to a simple balance linkage which

dispensed with the need for a cam at all. The hydraulic master cylinder was well tucked out of the way.

The frame was very substantial to withstand the loading it would be subject to. It followed normal BMW practice and, unlike its contemporary the Zundapp, which used pressings, was mainly tubular. It was of bolted construction and built up from four main members and four short bracing tubes with the top beam a substantial affair of head tube and formed sheet welded together. To its rear end was bolted the upper rear stay assembly comprising a tube formed to suit on each side and a cross tube. From the same attachment two small tubes forward and down to pick up on the crankcase, while a massive pressing attached to the back of the gearbox.

From the front of the top tube two smaller tubes dropped to support the front of the crankcase and bolted to the front end of the lower frame rails. These ran right back under the power unit to the rear wheel where they joined the upper rear stays and were crossbraced by a single tube.

It was a very substantial frame and well able to withstand the rigours of combat duty. No rear suspension was fitted as this would have complicated the rear wheel drive even further, but telescopic forks were used at the front. Most modern pictures show these fitted with gaiters, but contemporary ones indicate that the normal fork shrouds were used up to 1943 after which the gaiters were fitted.

The sidecar wheel was spring on a torsion bar mounted concentric with the drive shaft that ran across the chassis. The wheel moved in an arc round this shaft to maintain the drive gears in mesh and its total movement was restricted by a rubber cushioned stop on the chassis and two

Right **Line up of German traffic police and their BMW R51s probably pre-war**

Above **White wall tyres to match the fine uniform of this member of the Gendarmerie**

A celebration to mark 50 years of the *Zweizylinder-Maschinen*, **a favourite police motorcycle in over 100 countries in the world**

Two English policemen with their BMWs. From a small start many UK forces turned to the German flat twin. Two versions of a similar model with Avon fairings and Craven panniers

ears on the back of the gear housing.

Substantial mudguards were fitted with lifting handles on both, while the rear stay of the front one doubled as a stand for wheel changing. At the rear such needs were assisted by the mudguard hinging up to let the wheel roll out. The seating was typically German with a large saddle, pivoted at the nose and supported by a single vertically mounted coil spring. It was comfortable. The pillion passenger rode in state on his own seat supported by two springs with a grab handle just in front of the seat to hold onto. The entire pillion seat assembly could be removed to leave a carrier to supplement any panniers that may have been fitted.

The petrol tank held 24 litres and was fitted with a single tap with filter bowl. In the top of the tank was a toolbox as was common to many of the range at that time and it was on the lid of this that the desert-type air filter was mounted.

The electrical system included a battery mounted beside the gearbox on the left, while the headlamp carried the usual Bosch switch to control the lights with a speedometer behind it in the shell.

The sidecar was a solid functional single-seater built a little on the style of a trials or ISDT type. It was, however, a good deal heavier and much sturdier. A crash bar was attached to the chassis to protect the sidecar wheel, and in fact matched the one on the machine frame that kept the left cylinder from being knocked off by a passing Panzer. A good sized grab handle was bolted on the chair scuttle and there was no nonsense with screens or anything else likely to impede combat troops. A tonneau was supplied which did allow the men to carry their packs in the dry and the sidecar door was, like the tonneau, made in fabric so could be removed and stowed out of the way when the going was tough.

The chair body was pivot mounted at the front and suspended from a pair of quarter elliptic springs at the rear. It normally carried a spare wheel on the boot lid and had mountings for other fitments to suit its use and the locality. At the front on either side of the nose went further boxes which could be panniers, ammunition boxes or petrol containers of the type popularly known as 'jerricans' and holding over 20 litres of fuel. With two cans and a full tank this could give the outfit a range of around 800 km, or even more with careful driving. On the nose was a further mounting which could be used for a convoy light or machine gun depending on military duties.

It was a massive outfit and weighed in at 420 kg unladen, and a good deal more when fully loaded with men and equipment. It was good for 95 km/h while it would run as slowly as 3 km/h if required. The reverse gear was a very real need just to manoeuvre the outfit when parking it.

The R75 was used extensively by the German army all over Europe, in the snows of the Russian front and by the *Afrika Korps* across the desert. It

Another version of the R80/7 for the UK police, this one having twin front discs. RT fairing and flashing blue lamps

went up mountains and through mud and proved itself to be a very tough machine indeed.

It ran alongside the Zundapp which used the same sidecar wheel drive mechanism and was a flat twin but differed in a number of points. Unlike the BMW, it had a one-piece crankshaft with caged split needle roller big ends held together by cap screws. These could be changed by removing the opposing cylinder to gain access. Curiously the lubrication system was as on the BMW with spun discs collecting the oil centrifuged into them and thus feeding the big ends. The transmission side was nearly the same except that the Zundapp had a two-plate clutch but the frame was built up from pressings as were the forks which looked like telescopics but moved on links like girders. The frame surrounded the petrol tank and not much other than the light alloy cylinder heads protruded outside it. Like the BMW, it was very sturdy and very heavy and in 1944 both were dropped.

The production cost of either make was very high and both placed considerable emphasis on the need for good materials and close tolerance machining. In wartime the cost could be accepted but only if the machine gave optimum results, while once material shortages arose so did manufacturing problems. In operation the outfits could respond well to skilled driving but such are the vagaries of the sidecar that in the hands of the normal soldier they became a liability, not an asset. They failed to exploit the lightness and flexibility of use of the solo motorcycle and were thus limited in their fields of operation.

In the end the *Wehrmacht* followed the lead of the British army to lightweights and simple singles in the motorcycle field. The Allied armies had the benefit of the Jeep, while the Germans used the *Kübelwagen*. The Jeep was easy to handle, carried four and offered weather protection and something to crawl under during bombardment. They were cheap, simple and tough; went anywhere the R75 went and could be rolled over by a novice as easily as he could lift a sidecar wheel.

The German army gave up its sidecars, and with their departure went one of the most interesting of the BMW flat twins.

Away from war, the BMW was much used by German and other official government departments both before and after the war. In many cases the machine was attached to a sidecar of some form and early users included the fire fighting service which had an R12 coupled to a purpose-built box outfit. The German post office was another user with a box chair in pre-war days and a modified Isetta in postwar ones.

The most common user of the BMW motorcycle in its solo form has been the police. As would be expected they were a normal choice in Germany but have also been used by many other forces. Their appearance in Britain has already been mentioned but up and down Europe, on autostrada, autoroute, route nationale and elsewhere, with Police, *Polizei*, *Gendarmerie* and other variants on the front fairing, the appearance of a flat twin and a uniformed arm has so often meant 'pull over for a friendly chat about speed limits'.

A policeman on patrol duty spends a lot of time sitting on his machine and gets to know the riding position in general and the seat in particular very well. Their needs from a machine are much those of the tourer or sports tourer with the addition of a communication system, first-aid and fire gear, extra quiet for patrolling, good low speed handling for parades and the zip to head off a fast car. They must also handle the unexpected as one officer found when his pursuit of a trail bike took him and his R80/7 across a common. He caught the trail bike and mustered it back to the station cowboy style.

The BMW has proved itself to be very suitable for all these duties and to have the reliability to keep doing them day after day. It is for these reasons it is used by so many forces in so many countries throughout the world.

12 | Specials and replicas

There are not many BMW specials partly because of the sheer expense of the machines, but also because they are really too dignified for such work. On the other hand, there have been many replicas built by other makers, some close copies and others merely of similar form.

Such is the hold that the BMW has on the flat twin, shaft drive concept that any machine with that basic specification is automatically compared with the Munich model. The BMW itself is held by some to have been copied from the English ABC of 1919 and it in turn is supposed to have helped to inspire machines as diverse as the Zundapp, Velocette Valiant, Wooler, Condor and Gnome et Rhone.

Few of these were really at all like the BMW except in outline specification, but there have been several real replicas from both the Eastern block and the Far East. The latter, coming from Japan, were built in the early 1950s when the Japanese copied the European machines just as they copied one another. So Yamaha copied the 125 DKW just as did BSA and Harley-Davidson, Honda copied NSU and a firm called DSK or Daito Sangyo copied BMW. Unlike Harley-Davidson who built a rather crude copy of the R71 which failed to work as they did not have BMW experience of the flat twin, DSK copied the up to date R25 and R50. Rikuo also copied the single cylinder BMW and the German Embassy in Tokyo protested for it was hard to tell the two makes apart. Learning fast, the Japanese took no

Above left **The Zundapp flat twin. BMW in all but chain gearbox and needle roller big ends assembled to installed crankshaft**

Above right **The Neval Dnieper MT-12 combination of 750 cc with side valves, reverse gear and sidecar wheel drive as the R75.** *From Russia with love*

Below Motorcyclette Routière Type Lourd M53 **is how this BMW replica was described. In fact, a Russian machine from the Kiev Motor Factory about 1958 and of 500 cc**

Above **The Marusho Magnum with Honda type rear light, tank and forks**

Left above **500 cc Ratier built in Paris. Make started by assembling left over army spares and later evolved their own design in similar image**

Left **The Iso flat twin with electric starting and twin leading shoe front brake. The result of German/Italian collaborations that stemmed from Isetta work**

more notice than other countries before them protecting and expanding their own industries.

Innovation was rife in Japan at that time and copying, and then improving on the copy, was a fast way to learn. Even Honda built a flat twin then. Lilac were another to build an R50 copy, although theirs had forks like a BSA and a swinging fork at the rear end. Curiously Lilac also built a transverse vee twin very similar to the old German Victoria range and the more modern small Guzzi. Along with Lilac, DSK and Rikuo, Marusho and Tsubasa also built BMW-like machines with shaft drive, but in time these and nearly all the other 100 makers who had entered the 1950s with such high hopes went to the wall leaving just eight in 1965 and four names in the end. Their copying days were over.

In the East, or more precisely behind the Iron Curtain, the works at Eisenach fell to the Russians and from that situation came the EMW series of cars with badges similar to those from Munich but quartered in red and white with a star-like cross separating the colours. Under the guidance of the state controlled production agency, Autovelo, cars continued to be built but not to BMW standards. In time they reverted to the original Wartburg name used in the Eisenach works at the turn of the century.

On the motorcycle side this situation led to the production of EMW machines which were a direct copy of the R35 of 1937 with the horn moved and plunger rear suspension added. Such machines were built up to the middle 1950s along with the AWO, a 250 cc single very similar to the R25 and built at Suhl in Thuringia. These were made in the Simson car works and later adopted that name, while the works turned to two-strokes in the early 1960s. Until then the BMW style single continued and one was tested by *Motor Cycling* in August 1952. They commented on its likeness to the BMW and that the finish left something to be desired. The welding and castings in particular were poor and left unfinished but the paint and chrome were good.

The performance was limited but the handling good until a passenger was taken aboard. The second seat was mounted far to the rear, behind the rear wheel spindle and high up with a typical German style grab handle in front of it. This did nothing to help the steering. 45 mph cruising was thought to be reasonable with a 60 mph maximum.

In the 1970s came the Russian copies of the flat twin variously named Ural, Cossack and Neval. All were of 650 cc and based on the old pre-war R66 engine and frame. They did have the postwar swinging fork rear suspension as used on the R50

Above **Engine unit of the Marusho Magnum showing the very marked similarity to the Munich twin**

Below **Russian copy, in this case with the Ural name, based on a combination of pre and postwar models**

but the finish was distinctively agricultural. The Ural had many details identical to those of an early BMW but it lacked brakes, reasonable tyres and, early on, suffered from material selection and quality control in engine internals. This led to the adoption of various BMW parts in engine and transmission, and of European proprietary items in the areas of brakes and tyres. It suited those who enjoyed modifying a poorly made machine and who had the tools and experience to produce an acceptable counterfeit BMW.

In 1980 another version of the flat twin appeared from the same Russian stable — the Neval model MT-12 of 750 cc with side valves. It was a curious mixture of an R71 engine with the four-speed and reverse gearbox of the R75 together with the R75's sidecar wheel drive and differential. It only lacked the auxiliary two-speed box and the differential lock. It came with an sidecar permanently attached on the right — as with earlier Cossack chairs, this was based largely on the Wehrmacht type with crash bar, grab handle, spare wheel on the boot and tonneau cover, but no screen. Both sidecar and rear wheels had swinging arm suspension and all three wheels were interchangeable and fitted with 3·50 x 19 in. tyres.

In the realms of the special, the factory was one of the first to be involved when it was associated with the Iso concern. They built a 500 cc flat twin engine with ohv, electric start and shaft drive. They were tied up with BMW on the Isetta and required a suitable motorcycle chassis for their engine. The result was surprisingly like an R50. Later they built their own frame and fully enclosed the engine and gearbox unit on the lines of the later /5 series. The appearance of the final version was more like the Lilac but with Italian tank and very typical Italian rear suspension unit top supports.

Then there were the private owner jobs such as the fitting of an R90 engine into an R69 frame, or the more outstanding exercise with a Chevrolet Corvair flat 6 car engine in such a machine. Very neatly done too.

Less of a special and more a design exercise was the Booleroo, a machine devised by Trevor Innes who felt that motorcycle design was stagnant. He thought that machines should have a longer wheelbase so that rider and passenger could sit between the wheels and lower to reduce the height of the centre of gravity and give a more even fore and aft weight distribution. Anti-dive suspension and coupled brakes were other criteria and the low layout of the design made the choice of a BMW engine unit nearly automatic. The machine was built in 1980 and fitted with screen, 45-litre petrol tank, luggage compartment and paired headlights, all items to improve its touring and cruising abilities.

Finally there are the choppers. The BMW is an unlikely make for such purposes but there have been BMWs, both singles and twins, so modified and Krauser offered a mildly chopped flat twin in 1981 along with their four-valve head and out-and-out café racer.

In the USA Butler and Smith produced a very well executed chopper in 1973 using a R75/5 as the basis. Internally the engine was over-bored to 850 cc and fitted with pistons to give a compression ratio of 10·5:1 together with a number of other special parts. On the outside it had extended forks, twin front discs, twin headlights, a fat rear tyre, two-into-one exhaust, special lowered seat and a finish in Marano Pearl which changed its shade in sunlight. Various parts including the spokes, Bing bellmouths and fork springs were 24 carat gold-plated to give a touch of distinction to the finished machine.

It was a complete and instantly recognised chopper but with a discreet BMW style.

It was still—Beautifully Made Wonderbike, Best Motorcycle in the World, Berlin Motor Works, *Bayerische Motorrad Werke*, BMW.

Appendix

Specifications

Model	R24	R25	R25/3	R26	R27
Year from	**1949**	**1950**	**1953**	**1955**	**1960**
Year to	**1950**	**1953**	**1955**	**1960**	**1967**
Bore (mm)	68	68	68	68	68
Stroke (mm)	68	68	68	68	68
Capacity (cc)	247	247	247	247	247
Compression ratio (to 1)	6·75	6·5	7·0	7·5	8·2
Carburettor type	Bing	Bing	Bing	Bing	Bing
Carburettor size (mm)	22	22	24	26	26
Gear ratio: top	1·54	1·54	1·54	1·54	1·54
Gear ratio: 3rd	2·04	2·04	2·04	2·04	2·04
Gear ratio: 2nd	3·0	3·0	3·0	3·02	3·02
Gear ratio: 1st	6·1	6·1	6·1	5·33	5·33
Axle ratio: solo	4·18	4·5	4·16	4·16	4·16
Axle ratio: sidecar		5·2	4·8	5·2	5·20
Front tyre (in.)	3·00 × 19	3·25 × 19	3·25 × 18	3·25 × 18	3·25 × 18
Rear tyre (in.)	3·00 × 19	3·25 × 19	3·25 × 18	3·25 × 18	3·25 × 18
Brake front dia. (mm)				160	160
Brake front width (mm)				35	35
Brake rear dia. (mm)				160	160
Brake rear width (mm)				35	35
Front suspension	teles	teles	teles	Earles	Earles
Rear type	rigid	plunger	plunger	s/a	s/a
Petrol tank (litre)	12	12	12	15	15
Oil capacity (litre)	1·5	1·25	1·25	1·25	1·25
Ignition system	coil	coil	magneto	magneto	coil
Generator type	dynamo	dynamo	dynamo	dynamo	dynamo
Output (watts)	45	45		60	60
Battery (volts)	6	6	6	6	6
Wheelbase (mm)		1370			
Ground clearance (mm)		125			
Seat height (mm)	710	710	730	770	770
Width (mm)	750	750	760	660	660
Length (mm)	2020	2020	2065	2090	2090
Dry weight (lb/kg)	287/130	309/140	331/150	348/158	357/162

Model	R24	R25	R25/3	R26	R27
Year from	1949	1950	1953	1955	1960
Year to	1950	1953	1955	1960	1967
Power: bhp	12	12	13	15	18
@ rpm	5600	5600	5800	6400	7400

Model	R51/2	R51/3	R50	R50S
Year from	1950	1951	1955	1960
Year to	1950	1954	1969	1963
Bore (mm)	68	68	68	68
Stroke (mm)	68	68	68	68
Capacity (cc)	494	494	494	494
Compression ratio (to 1)	6·3	6·3	6·8	9·2
inlet opens BTDC			6	
inlet closes ABDC			34	
exhaust opens BBDC			34	
exhaust closes ATDC			6	
Inlet clearance (mm)			0·15	0·15
Exhaust clearance (mm)			0·20	0·20
Ignition timing °			39	39
Points gap (mm)	0·35	0·35	0·35	0·35
Carburettor type	Bing	Bing	Bing	Bing
Carburettor size	22	22	24	26
Gear ratio: top	1·3	1·3	1·54	1·54
Gear ratio: 3rd	1·7	1·7	2·04	1·938
Gear ratio: 2nd	2·28	2·28	3·02	2·725
Gear ratio: 1st	3·6	3·6	5·33	4·171
Axle ratio: solo	3·89	3·89	3·18	3·58
Axle ratio: sidecar	4·62	4·57	4·25	4·33
Front tyre (in.)	3·50 × 19	3·50 × 19	3·50 × 18	3·50 × 18
Rear tyre (in.)	3·50 × 19	3·50 × 19	3·50 × 18	3·50 × 18
Brake front dia. (mm)	200	200	200	200
Brake front width (mm)			35	
Brake rear dia. (mm)	200	200	200	200
Brake rear width (mm)			35	
Front suspension	teles	teles	Earles	Earles
Rear type	plunger	plunger	s/a	s/a
Spring movement (mm)			105	105
Petrol tank (litre)	14	17	17	17
Oil capacity (litre)	2	2	2	2
Box capacity (litre)			0·8	0·8
Ignition system	coil	magneto	magneto	magneto
Generator type	dynamo	dynamo	dynamo 1	dynamo
Output (watts)	60	60	60 2	60
Battery (volts)	6	6	6 2	6
Wheelbase (mm)			1422	
Ground clearance (mm)			135	
Seat height (mm)	720	720	725	725

Model	R51/2	R51/3	R50	R50S
Year from	**1950**	**1951**	**1955**	**1960**
Year to	**1950**	**1954**	**1969**	**1963**
Width (mm)	815	790	660	660
Length (mm)	2130	2130	2125	2125
Dry weight (lb/kg)	408/185	419/190	430/195	437/198
Power: bhp	24	24	26	35
@ rpm	5800	5800	5800	7650

1 1969—alternator   **2** 1968—12 volt, 100 watt

Model	R67	R68	R69	R60	R69S
Year from	**1951**	**1952**	**1955**	**1956**	**1960**
Year to	**1956**	**1954**	**1960**	**1967**	**1969**
Bore (mm)	72	72	72	72	72
Stroke (mm)	73	73	73	73	73
Capacity (cc)	594	594	594	594	594
Compression ratio (to 1)	5·6	7·5–7·7	8·0	6·5 **1**	9·5
Valve timing:					
inlet opens BTDC				6	4
inlet closes ABDC				34	44
exhaust opens BBDC				34	44
exhaust closes ATDC				6	4
Inlet clearance (mm)			0·15	0·15	0·15
Exhaust clearance (mm)			0·20	0·20	0·20
Ignition timing °			39	39	39
Points gap (mm)	0·35	0·35	0·35	0·35	0·35
Carburettor type	Bing	Bing	Bing	Bing	Bing
Carburettor size (mm)	24	26	26	24	26
Gear ratio: top	1·3	1·3	1·54	1·54	1·54
Gear ratio: 3rd	1·7	1·7	2·04	2·04	1·938
Gear ratio: 2nd	2·28	2·28	3·02	3·02	2·725
Gear ratio: 1st	3·6	4·0	5·33	5·33	4·171
Axle ratio: solo	3·56	3·89	3·18	2·91	3·13
Axle ratio: sidecar	4·38		4·25	3·86	4·33
Front tyre (in.)	3·50 × 19	3·50 × 19	3·50 × 18	3·50 × 18	3·50 × 18
Rear tyre (in.)	3·50 × 19	3·50 × 19	3·50 × 18	3·50 × 18 **2**	3·50 × 18
Brake front dia. (mm)	200	200	200	200	200
Brake front width (mm)				35	35
Brake rear dia. (mm)	200	200	200	200	200
Brake rear width (mm)				35	35
Front suspension	teles	teles	Earles	Earles	Earles
Rear type	plunger	plunger	s/a	s/a	s/a
Spring movement (mm)			105	105	105
Petrol tank (litre)	17	17	17	17	17
Oil capacity (litre)	2	2	2	2	2
Box capacity (litre)			0·8	0·8	0·8
Ignition system	magneto	magneto	magneto	magneto	magneto

Model	R67	R68	R69	R60	R69S
Year from	1951	1952	1955	1956	1960
Year to	1956	1954	1960	1967	1969
Generator type	dynamo	dynamo	dynamo	dynamo	dynamo 3
Output (watts)	60	60	60	60	60 4
Battery (volts)	6	6	6	6	6 4
Wheelbase (mm)			1416	1422	1422
Ground clearance (mm)			127	135	135
Seat height (mm)	720	725	725	725	725
Width (mm)	790	790	722	660	722
Length (mm)	2130	2150	2125	2125	2125
Dry weight (lb/kg)	423/192	425/193	445/202	430/195	445/202
Power: bhp	26 5	35	35	28 6	42
@ rpm	5500 5	7000	6800	5600 6	7000

1 1960—7·5   **2** sidecar spec.—4·00 × 18   **3** 1969—alternator   **4** 1968—12 volt, 100 watt   **5** 1952—28/5600
6 1960—30/5800

Model	R50/5	R60/5	R60/6	R60/7
Year from	1969	1969	1973	1976
Year to	1973	1973	1976	1978
Bore (mm)	67	73·5	73·5	73·5
Stroke (mm)	70·6	70·6	70·6	70·6
Capacity (cc)	498	599	599	599
Inlet valve dia. (mm)	34	38	38	38
Exhaust valve dia. (mm)	32	34	34	34
Compression ratio (to 1)	8·6	9·2	9·2	9·2
Valve timing:				
inlet opens BTDC	TDC	TDC	TDC	TDC 1
inlet closes ABDC	40	40	40	40
exhaust opens BBDC	40	40	40	40
exhaust closes ATDC	TDC	TDC	TDC	TDC
Inlet clearance (mm)	0·15	0·15	0·15	0·10
Exhaust clearance (mm)	0·20	0·20	0·20	0·15
Ignition timing °	34	34	34	31
Points gap (mm)	0·35	0·35	0·35	0·35
Carburettor type	Bing	Bing	Bing	Bing
Carburettor size	26	26	26	26
Gear ratio: top	1·50	1·50	1·50	1·50
Gear ratio: 4th			1·67	1·67
Gear ratio: 3rd	1·875	1·875	2·07	2·07
Gear ratio: 2nd	2·578	2·578	2·86	2·86
Gear ratio: 1st	3·896	3·896	4·40	4·40
Axle ratio: solo	3·56	3·36	3·36	3·36
Axle ratio: option			3·56	3·56
Axle ratio: USA				3·56
Front tyre (in.)	3·25 × 19	3·25 × 19	3·25 × 19	3·25 × 19
Rear tyre (in.)	4·00 × 18	4·00 × 18	4·00 × 18	4·00 × 18

Model	R50/5	R60/5	R60/6	R60/7
Year from	**1969**	**1969**	**1973**	**1976**
Year to	**1973**	**1973**	**1976**	**1978**
Brake front dia. (mm)	200	200	200	260 disc
Brake front width (mm)	30	30	30	
Brake rear dia. (mm)	200	200	200	200
Brake rear width (mm)	30	30	30	30
Front suspension	teles	teles	teles	teles
Spring movement (mm)	208	208	208	208
Rear type	s/a	s/a	s/a	s/a
Spring movement (mm)	125	125	125	125
Petrol tank (litre)	22 **2**	22 **2**	18	24
Oil capacity (litre)	2	2	2	2
Box capacity (litre)	0·8	0·8	0·8	0·8
Ignition system	coil	coil	coil	coil
Generator type	alternator	alternator	alternator	alternator
Output (watts)	180	180	280	280
Battery (volts)	12	12	12	12
Wheelbase (mm)	1385 **3**	1385 **3**	1465	1465
Ground clearance (mm)	165	165	165	165
Seat height (mm)	850	850	810	810
Width (mm)	740	740	740	746
Length (mm)	2100	2100	2180	2130
Dry weight (lb/kg)	408/185	419/190	441/200	430/195
Wet weight (lb/kg)	441/200	452/205	464/210	474/215
Power: bhp	32	40	40	40
@ rpm	6400	6400	6400	6400

1 1978—6/34/46/6   **2** 1972—17·5   **3** 1973—1465

Model	R75/5	R75/6	R75/7	R80/7
Year from	**1969**	**1973**	**1976**	**1977**
Year to	**1973**	**1976**	**1977**	**1980**
Bore (mm)	82	82	82	84·8
Stroke (mm)	70·6	70·6	70·6	70·6
Capacity (cc)	746	746	746	797
Inlet valve dia. (mm)	42	42	42	42
Exhaust valve dia. (mm)	38	38	38	38
Compression ratio (to 1)	9·0	9·0	9·0	9·2 **1**
Valve timing:				

Model	R75/5	R75/6	R75/7	R80/7
Year from	**1969**	**1973**	**1976**	**1977**
Year to	**1973**	**1976**	**1977**	**1980**
inlet opens BTDC	10	10	10	10 **2**
inlet closes ABDC	50	50	50	50
exhaust opens BBDC	50	50	50	50
exhaust closes ATDC	10	10	10	10
Inlet clearance (mm)	0·15	0·15	0·10	0·10
Exhaust clearance (mm)	0·20	0·20	0·15	0·15
Ignition timing °	34	34	31	31 **3**
Points gap (mm)	0·35	0·35	0·35	0·35 **3**
Carburettor type	Bing	Bing	Bing	Bing
Carburettor size	32	32	32	32
Gear ratio: top	1·50	1·50	1·50	1·50
Gear ratio: 4th		1·67	1·67	1·67
Gear ratio: 3rd	1·875	2·07	2·07	2·07
Gear ratio: 2nd	2·578	2·86	2·86	2·86
Gear ratio: 1st	3·896	4·40	4·40	4·40
Axle ratio: solo	2·91 **4**	3·20	3·20	3·20 **5**
Axle ratio: option		3·36		3·36
Axle ratio: USA			3·36	3·36
Front tyre (in.)	3·25 × 19	3·25 × 19	3·25 × 19	3·25 × 19
Rear tyre (in.)	4·00 × 18	4·00 × 18	4·00 × 18	4·00 × 18
Brake front dia. (mm)	200	260 disc	260 disc	260 disc **6**
Brake front width (mm)	30			
Brake rear dia. (mm)	200	200	200	200
Brake rear width (mm)	30	30	30	30
Front suspension	teles	teles	teles	teles
Spring movement (mm)	208	208	208	208
Rear type	s/a	s/a	s/a	s/a
Spring movement (mm)	125	125	125	125
Petrol tank (litre)	22 **7**	18	24	24
Oil capacity (litre)	2	2	2	2
Box capacity (litre)	0·8	0·8	0·8	0·8
Ignition system	coil	coil	coil	coil
Generator type	alternator	alternator	alternator	alternator
Output (watts)	180	280	280	280
Battery (volts)	12	12	12	12
Wheelbase (mm)	1385 **8**	1465	1465	1465
Ground clearance (mm)	165	165	165	165
Seat height (mm)	850	810	810	810
Width (mm)	740	740	746	746
Length (mm)	2100	2180	2130	2180
Dry weight (lb/kg)	419/190	441/200	430/195	430/195
Wet weight (lb/kg)	452/205	464/210	474/215	474/215

Model	**R75/5**	**R75/6**	**R75/7**	**R80/7**
Year from	**1969**	**1973**	**1976**	**1977**
Year to	**1973**	**1976**	**1977**	**1980**
Power: bhp	50	50	50	55 **9**
@ rpm	6200	6200	6200	7000 **9**

1 8·0 option   **2** 1978—16/44/56/4   **3** 1979—32/0·45   **4** 3·21 option   **5** 3·36 option   **6** 1978—twin disc
7 1972—17·5   **8** 1973—1465   **9** 50/7250 option

Model	**R90/6**	**R90/S**	**R100/7**	**R100S**
Year from	**1973**	**1973**	**1976**	**1976**
Year to	**1976**	**1976**	**1978**	**1980**
Bore (mm)	90	90	94	94
Stroke (mm)	70·6	70·6	70·6	70·6
Capacity (cc)	898	898	980	980
Inlet valve dia. (mm)	42	42	42	44
Exhaust valve dia. (mm)	40	40	40	40
Compression ratio (to 1)	9·0	9·5	9·0	9·5
Valve timing:				
inlet opens BTDC	10	10	10 **1**	10 **1**
inlet closes ABDC	50	50	50	50
exhaust opens BBDC	50	50	50	50
exhaust closes ATDC	10	10	10	10
Inlet clearance (mm)	0·15	0·15	0·10	0·10
Exhaust clearance (mm)	0·20	0·20	0·15	0·15
Ignition timing °	34	34	31	31 **2**
Points gap (mm)	0·35	0·35	0·35	0·35 **2**
Carburettor type	Bing	Dell'Orto	Bing	Bing
Carburettor size	32	38	32	40
Gear ratio: Top	1·50	1·50	1·50	1·50
Gear ratio: 4th	1·67	1·67	1·67	1·67
Gear ratio: 3rd	2·07	2·07	2·07	2·07
Gear ratio: 2nd	2·86	2·86	2·86	2·86
Gear ratio: 1st	4·40	4·40	4·40	4·40
Axle ratio: solo	3·09	3·00	3·09	2·91
Axle ratio: option	3·20	2·91	3·20	3·00
Axle ratio: USA			3·00	2·91
Front tyre (in.)	3·25 × 19	3·25 × 19	3·25 × 19	3·25 × 19
Rear tyre (in.)	4·00 × 18	4·00 × 18	4·00 × 18	4·00 × 18
Brake front dia. (mm)	260 disc	260 twin disc	260 disc **3**	260 twin disc
Brake rear dia. (mm)	200	200	200	200 **4**
Brake rear width (mm)	30	30	30	30
Front suspension	teles	teles	teles	teles
Spring movement (mm)	208	208	208	208
Rear type	s/a	s/a	s/a	s/a
Spring movement (mm)	125	125	125	125
Petrol tank (litre)	18	24	24	24

Model	R90/6	R90/S	R100/7	R100S
Year from	1973	1973	1976	1976
Year to	1976	1976	1978	1980
Oil capacity (litre)	2	2	2	2
Box capacity (litre)	0.8	0.8	0.8	0.8
Ignition system	coil	coil	coil	coil
Generator type	alternator	alternator	alternator	alternator
Output (watts)	280	280	280	280
Battery (volts)	12	12	12	12
Wheelbase (mm)	1465	1465	1465	1465
Ground clearance (mm)	165	165	165	165
Seat height (mm)	810	810	810	820
Width (mm)	740	740	746	746
Length (mm)	2180	2180	2130	2120
Dry weight (lb/kg)	441/200	452/205	430/195	441/200
Wet weight (lb/kg)	464/210	474/215	474/215	485/220
Power: bhp	60	67	60	65
@ rpm	6500	7000	6500	6600

1 1976—16/44/56/4 **2** 1979—32/0.45 **3** 1978—twin disc **4** 1978—260 disc **5** 1978—70/7250

Model	R100RS	R100RT	R100T	R100	R100CS
Year from	1976	1976	1979	1981	1981
Year to	1984	1984	1980	1983	1983
Bore (mm)	94	94	94	94	94
Stroke (mm)	70.6	70.6	70.6	70.6	70.6
Capacity (cc)	980	980	980	980	980
Inlet valve dia. (mm)	44	44	44	44	44
Exhaust valve dia (mm)	40	40	40	40	40
Compression ratio (to 1)	9.5	9.5	9.5	8.2	9.5
Valve timing:					
Inlet opens BTDC	10 **1**	16	16	16	16
Inlet closes ABDC	50	44	44	44	44
Exhaust opens BBDC	50	56	56	56	56
Exhaust closes ATDC	10	4	4	4	4
Inlet clearance (mm)	0.10	0.10	0.10	0.10	0.10
Exhaust clearance (mm)	0.15	0.15	0.15	0.15	0.15
Ignition timing °	31 **2**	32	32	32	32
Points gap (mm)	0.35 **3**	0.45	0.45		
Carburettor type	Bing	Bing	Bing	Bing	Bing
Carburettor size	40	40	40	40	40
Gear ratio: top	1.50	1.50	1.50	1.50	1.50
Gear ratio: 4th	1.67	1.67	1.67	1.67	1.67
Gear ratio: 3rd	2.07	2.07	2.07	2.07	2.07
Gear ratio: 2nd	2.86	2.86	2.86	2.86	2.86
Gear ratio: 1st	4.40	4.40	4.40	4.40	4.40
Axle ratio: solo	3.00	3.00	3.00	3.00	2.91

Model	R100RS	R100RT	R100T	R100	R100CS
Year from	**1979**	**1976**	**1979**	**1981**	**1981**
Year to	**1984**	**1984**	**1980**	**1983**	**1983**
Axle ratio: option	2.91	2.91	2.91		
Axle ratio: USA	2.91				
Front tyre (in.)	3.25 x 19	3.25 x 19	3.25 x 19	3.25 x 19	3.25 x 19
Rear tyre (in.)	4.00 x 18	4.00 x 18	4.00 x 18	4.00 x 18	4.00 x 18
Brake front dia. (mm)	260 twin disc	260 twin disc	260 twin disc	260 twin disc	260 twin disc
Brake rear dia. (mm)	200 **4**	260 disc	200	200	200
Brake rear width (mm)	30		30	30	30
Front suspension	teles	teles	teles	teles	teles
Spring movement (mm)	208	208	208	200	200
Rear type	s/a	s/a	s/a	s/a	s/a
Spring movement (mm)	125	125	125	125	125
Petrol tank (litre)	24	24	24	24	24
Oil capacity (litre)	2	2	2	2	2
Box capacity (litre)	0.8	0.8	0.8	0.8	0.8
Ignition system	coil **5**	coil **5**	coil	electronic	electronic
Generator type	alternator	alternator	alternator	alternator	alternator
Output (watts)	280	280	280	280	280
Battery (volts)	12	12	12	12	12
Wheelbase (mm)	1465	1465	1465	1465	1465
Ground clearance (mm)	165	165	165		
Seat height (mm)	820	820	820	820	820
Width (mm)	746	746	746	746	746
Length (mm)	2210	2210	2210	2210	2210
Dry weight (lb/kg)	463/210	472/214	437/198	437/198	441/200
Wet weight (lb/kg)	507/230	516/234	481/218	481/218	485/220
Power: bhp	70	70	65	67	70
@ rpm	7250 **6**	7250 **6**	6600	7000	7000

1 1978—16/44/56/4 **2** 1979—32 **3** 1979—0.45 **4** 1977—260 disc **5** 1981—electronic **6** 1981—7000

Model	R45	R65	R65LS	R80G/S
Year from	**1978**	**1978**	**1982**	**1981**
Year to	**1985**	**1985**	**1985**	**1987**
Bore (mm)	70	70	82	84.8
Stroke (mm)	61.5	61.5	61.5	70.6
Capacity (cc)	473	650	650	797
Inlet valve dia. (mm)	34	38 **1**	40	42
Exhaust valve dia (mm)	32	34 **9**	34 **9**	38
Compression ratio (to 1)	9.2 **2**	9.2	9.2	8.2
Valve timing:				
Inlet opens BTDC	16	16	16	
Inlet closes ABDC	44	44	44	
Exhaust opens BBDC	56	56	56	
Exhaust closes ATDC	4	4	4	
Inlet clearance (mm)	0.10	0.10	0.10	0.10

Model	R45	R65	R65LS	R80G/S
Year from	**1978**	**1978**	**1982**	**1981**
Year to	**1985**	**1985**	**1985**	**1987**
Exhaust clearance (mm)	0.20	0.20	0.20	0.15
Ignition timing °	32	32	32	32
Points gap (mm)	0.45	0.45		
Carburettor type	Bing	Bing	Bing	Bing
Carburettor size	28 **3**	32	32	32
Gear ratio: top	1.50	1.50	1.50	1.50
Gear ratio: 4th	1.67	1.67	1.67	1.67
Gear ratio: 3rd	2.07	2.07	2.07	2.07
Gear ratio: 2nd	2.86	2.86	2.86	2.86
Gear ratio: 1st	4.40	4.40	4.40	4.40
Axle ratio: solo	3.89	3.44	3.89	3.36
Axle ratio: option	4.25		4.25	
Front tyre (in.)	3.25 x 18	3.25 x 18	3.25 x 18	3.00 x 21
Rear tyre (in.)	4.00 x 18	4.00 x 18	4.00 x 18	4.00 x 18
Brake front dia. (mm)	260 disc	260 disc	260 twin disc	260 disc
Brake rear dia. (mm)	200	200	220	200
Brake rear width (mm)	30	30	30	30
Front suspension	teles	teles	teles	teles
Spring movement (mm)	175	175	175	200
Rear tyre	s/a	s/a	s/a	s/a
Spring movement (mm)	110	110	110	170
Petrol tank (litre)	22	22	22	19.5 **10**
Oil capacity (litre)	2	2	2	2
Box capacity (litre)	0.8	0.8	0.8	0.8
Ignition system	coli **4**	coli **4**	electronic	electronic
Generator type	alternator	alternator	alternator	alternator
Output (watts)	280	280	280	280
Battery (volts)	12	12	12	12
Wheelbase (mm)	1390 **5**	1390 **5**	1400	1447
Ground clearance (mm)	105	105		218
Seat height (mm)	770 **6**	770 **6**	810	860 **11**
Width (mm)	688	688	688	820
Length (mm)	2110	2110	2110	2230
Dry weight (lb/kg)	408/185	408/185	408/185	368/167
Wet weight (lb/kg)	452/205	452/205	452/205	410/186
Power: bhp	35 **7**	45 **8**	50	50
@ rpm	7250 **7**	7250	7250 **7**	6500

1 1980—40 **2** 8.2 option **3** 26 option **4** 1981—electronic **5** 1981—1400 **6** 1981—810
7 27/6500 optio **8** 1981—50 **9** 1985—36 **10** Paris-Dakar—32 **11** 1985 Paris-Dakar—845

Model	R80ST	R80RT	R80RT	R80	R65
Year from	**1982**	**1982**	**1985**	**1985**	**1986**
Year to	**1984**	**1984**	**1994**	**1994**	**1991**
Bore (mm)	84.8	84.8	84.8	84.8	82
Stroke (mm)	70.6	70.6	70.6	70.6	61.5
Capacity (cc)	797	797	797	797	650
Inlet valve dia. (mm)	42	42	42	42	40 **1**
Exhaust valve dia (mm)	38	38	38	38	36 **2**
Compression ratio (to 1)	8.2	8.2	8.2	8.2	8.7 **3**
Carburettor type	Bing	Bing	Bing	Bing	Bing
Carburettor size	32	32	32	32	32 **4**
Gear ratio: top	1.50	1.50	1.50	1.50	1.50
Gear ratio: 4th	1.67	1.67	1.67	1.67	1.67
Gear ratio: 3rd	2.07	2.07	2.07	2.07	2.07
Gear ratio: 2nd	2.86	2.86	2.86	2.86	2.86
Gear ratio: 1st	4.40	4.40	4.40	4.40	4.40
Axle ratio: solo	3.36	3.36	3.36 **5**	3.20	3.36 **6**
Front tyre (in.)	100/90H19	3.25x19	90/90H18	90/90H18	90/90H18
Rear tyre (in.)	120/90H18	4.00x18	120/90H18	120/90H18	120/90H18
Brake front dia (mm)	260 disc	260 twin disc	285 disc **7**	285 disc	285 disc
Brake rear dia. (mm)	200	200	200	200	200
Front suspension	teles	teles	teles	teles	teles
Spring movement (mm)	175	200	185 **8**	185 **8**	175
Rear type	s/a	s/a	s/a	s/a	s/a
Spring movement (mm)	153	125	121	121	121
Petrol tank (litre)	19	24	22	22	22
Ignition system	electronic	electronic	electronic	electronic	electronic
Generator type	alternator	alternator	alternator	alternator	alternator
Output (watts)	280	280	280 **9**	280 **9**	280
Battery (volts)	12	12	12	12	12
Wheelbase (mm)	1446	1465	1447	1447	1447
Seat height (mm)	845	820	807	807	807
Width (mm)	746	746	960	800	800
Length (mm)	2180	2220	2175	2175	2175
Wet weight (lb/kg)	437/198	516/234	500/227	463/210	452/205
Power: bhp	50	50	50	50 **10**	48 **11**
@ rpm	6500	6500	6500	6500 **12**	7250 **13**

1 1989—34 **2** 1989—32 **3** 1989—8.4 **4** 1989—26 **5** 1987—3.20, 1991—3.36 **6** 1989—3.20 **7** 1990—285 twin disc **8** 1986—175 **9** 1992—240 **10** 1994—34 **11** 1989—27 **12** 1994—6000 **13** 1989—5500

Model	R100RS	R100RT	R80GS	R100GS	R65GS
Year from	**1987**	**1988**	**1988**	**1988 1**	**1988**
Year to	**1992**	**1995**	**1994**	**1994 1**	**1990**
Bore (mm)	94	94	84.8	94	82
Stroke (mm)	70.6	70.6	70.6	70.6	61.5
Capacity (cc)	980	980	797	980	650
Inlet valve dia. (mm)	42	42	42	42	34

Model	R100RS	R100RT	R80GS	R100GS	R65GS
Exhaust valve dia (mm)	40	40	40	40	32
Compression ratio (to 1)	8.45	8.45	8.2	8.5	8.4
Carburettor type	Bing	Bing	Bing	Bing	Bing
Carburettor size	32	32	32	40	26
Gear ratio: top	1.50	1.50	1.50	1.50	1.50
Gear ratio: 4th	1.67	1.67	1.67	1.67	1.67
Gear ratio: 3rd	2.07	2.07	2.07	2.07	2.07
Gear ratio: 2nd	2.86	2.86	2.86	2.86	2.86
Gear ratio: 1st	4.40	4.40	4.40	4.40	4.40
Axle ratio: solo	3.00	3.09 **2**	3.09 **3**	3.20 **4**	3.44
Front tyre (in.)	90/90H18	90/90H18	90/90T21	90/90T21	3.00x21
Rear tyre (in.)	120/90H18	120/90H18	130/80T17	130/80T17	4.00x18
Brake front dia (mm)	285 twin disc	285 twin disc	285 disc	280 disc **5**	260 disc
Brake rear dia. (mm)	200	200	200	200	200
Front suspension	teles	teles	teles	teles	teles
Spring movement (mm)	175	175	225	225	200
Rear type	s/a	s/a	Paralever	Paralever	s/a
Spring movement (mm)	121	121	180	180	170
Petrol tank (litre)	22	22	26 **6**	26 **7**	19.5
Ignition system	electronic	electronic	electronic	electronic	electronic
Generator type	alternator	alternator	alternator	alternator	alternator
Output (watts)	280 **8**	280 **8**	280 **8**	280 **8**	280
Battery (volts)	12	12	12	12	12
Wheelbase (mm)	1447	1447	1513	1513	1465
Seat height (mm)	807	807	850	850	860
Width (mm)	800	960	1000	1000	1000
Length (mm)	2175	2175	2290	2290	2230
Wet weight (lb/kg)	505/229	516/234	463/210 **9**	463/210 **10**	436/198
Power: bhp	60	60	50	60	27
@ rpm	6500	6500	6500	6500	5500

1 Paris-Dakar—1989-1995 **2** 1991—3.00 **3** 1991—3.20 **4** 1991—3.09 **5** 1993—285 **6** 1991—24
7 1991—24, Paris-Dakar—35 **8** 1992—240 **9** 1991—474/215 **10** 1991—485/220, Paris-Dakar—520/236

Model	R100R	R80R	R1100RS	R1100GS	R1100R
Year from	**1992**	**1993**	**1993**	**1994**	**1995**
Year to	**1995**	**1994**	**1997**	**1997**	**1997**
Bore (mm)	94	84.8	99	99	99
Stroke (mm)	70.6	70.6	70.5	70.5	70.5
Capacity (cc)	980	797	1085	1085	1085
Inlet valve dia. (mm)	42	42	36	36	36
Exhaust valve dia (mm)	40	40	31	31	31
Compression ratio (to 1)	8.5	8.2	10.7	10.3	10.3
Carburettor type	Bing	Bing	n/a	n/a	n/a
Carburettor size	40	32	n/a	n/a	n/a
Gear ratio: top	1.50	1.50	1.45	1.45	1.45
Gear ratio: 4th	1.67	1.67	1.74	1.74	1.74

Model	R100R	R80R	R1100RS	R1100GS	R1100R
Gear ratio: 3rd	2.07	2.07	2.13	2.13	2.13
Gear ratio: 2nd	2.86	2.86	2.91	2.91	2.91
Gear ratio: 1st	4.40	4.40	4.16	4.16	4.16
Axle ratio: solo	3.09	3.20	2.81	3.00	3.00
Front tyre (in.)	110/80V18	110/80V18	120/70ZR17	110/80H19	120/70ZR17
Rear tyre (in.)	140/80V17	140/80V17	160/60ZR18	150/70H17	160/60ZR18
Brake front dia (mm)	285 disc **1**	285 disc	305 twin disc	305 twin disc	305 twin disc
Brake rear dia. (mm)	200	200	285 disc	276 disc	276 disc
Front suspension	teles	teles	Telelever	Telelever	Telelever
Spring movement (mm)	135	135	120	190	120
Rear type	Paralever	Paralever	Paralever	Paralever	Paralever
Spring movement (mm)	140	140	135	200	135
Petrol tank (litre)	24	24	23	25	21
Ignition system	electronic	electronic	electronic	electronic	electronic
Generator type	alternator	alternator	alternator	alternator	alternator
Output (watts)	240	240	700	700	700
Battery (volts)	12	12	12	12	12
Wheelbase (mm)	1513	1513	1473	1499 **2**	1487
Seat height (mm)	800	800	780 **3**	840 **4**	760 **5**
Width (mm)	1000	1000	920	920	898
Length (mm)	2210	2210	2175	2196	2197
Wet weight (lb/kg)	481/218	478/217	527/239	536/243	518/235
Power: bhp	60	50	90	80	80
@ rpm	6500	6500	7250	6750	6750

1 1995—285 twin disc **2** 1995—1509 **3** or 800 or 820 **4** or 860 **5** or 780 or 800

Model	R850R	R1100RT	K100	K100RS	K100RT
Year from	**1995**	**1996**	**1984**	**1984**	**1984**
Year to	**1997**	**1997**	**1990**	**1989**	**1988**
Bore (mm)	87.5	99	67	67	67
Stroke (mm)	70.5	70.5	70	70	70
Capacity (cc)	848	1085	987	987	987
Inlet valve dia. (mm)	32	36	34	34	34
Exhaust valve dia (mm)	27	31	28 **1**	28 **1**	28 **1**
Compression ratio (to 1)	10.3	10.7	10.2	10.2	10.2
Gear ratio: top	1.45	1.45	1.67	1.67	1.67
Gear ratio: 4th	1.74	1.74	1.88	1.88	1.88
Gear ratio: 3rd	2.13	2.13	2.30	2.30	2.30
Gear ratio: 2nd	2.91	2.91	2.96	2.96	2.96
Gear ratio: 1st	4.16	4.16	4.50	4.50	4.50
Axle ratio: solo	3.36	2.91	2.91	2.81	2.91
Front tyre (in.)	120/70ZR17	120/70ZR17	100/90V18	100/90V18	100/90V18
Rear tyre (in.)	160/60ZR18	160/60ZR18	130/90V17	130/90V17	130/90V17
Brake front dia (mm)	305 twin disc	305 twin disc	285 twin disc	285 twin disc	285 twin disc
Brake rear dia. (mm)	276 disc	276 disc	285 disc	285 disc	285 disc
Front suspension	Telelever	Telelever	teles	teles	teles
Spring movement (mm)	120	120	185	185	185

Model	R850R	R1100RT	K100	K100RS	K100RT
Rear type	Paralever	Paralever	s/a	s/a	s/a
Spring movement (mm)	135	135	110	110	110
Petrol tank (litre)	21	26	22 **2**	22	22
Ignition system	electronic	electronic	electronic	electronic	electronic
Generator type	alternator	alternator	alternator	alternator	alternator
Output (watts)	700	700	460	460	460
Battery (volts)	12	12	12	12	12
Wheelbase (mm)	1487	1485	1516 **3**	1516	1516
Seat height (mm)	760 **4**	780 **5**	810 **6**	810	810
Width (mm)	898	898	960	800	916
Length (mm)	2197	2195	2200 **7**	2200 **7**	2200 **7**
Wet weight (lb/kg)	518/235	622/282	527/239	549/249 **8**	558/253 **9**
Power: bhp	70 **10**	90	90	90	90
@ rpm	7000	7250	8000	8000	8000

1 1987—30 **2** 1988—21 **3** 1990—1511 **4** or 780 or 800 **5** or 800 or 820 **6** 1988—760 or 800
7 1985—2220 **8** 1985—558/253 **9** 1985—580/263 **10** or 34

Model	K75C	K75S	K75	K100LT	K1	K75RT
Year from	**1986**	**1986**	**1987**	**1987**	**1989**	**1990**
Year to	**1988**	**1995**	**1996**	**1991**	**1993**	**1996**
Bore (mm)	67	67	70	67	67	67
Stroke (mm)	70	67	70	70	70	70
Capacity (cc)	740	740	740	987	987	740
Inlet valve dia. (mm)	34	34	34	34	26.5	34
Exhaust valve dia (mm)	28 **1**	28 **1**	30	30	23	30
Compression ratio (to 1)	11.0	11.0	11.0	10.2	11.0	11.0
Gear ratio: top	1.67	1.67	1.67	1.67	1.61	1.67
Gear ratio: 4th	1.88	1.88	1.88	1.88	1.88	1.88
Gear ratio: 3rd	2.30	2.30	2.30	2.30	2.30	2.30
Gear ratio: 2nd	2.96	2.96	2.96	2.96	2.96	2.96
Gear ratio: 1st	4.50	4.50	4.50	4.50	4.50	4.50
Axle ratio: solo	3.20	3.20	3.20	2.91	2.75 **2**	3.20
Front tyre (in.)	100/90H18	100/90V18	100/90H18	100/90V18 **3**	120/70V17	100/90V18
Rear tyre (in.)	120/90H18	130/90V17	120/90H18 **4**	130/90V17 **5**	160/60V18	130/90V17
Brake front dia (mm)	285 twin disc	285 twin disc	285 twin disc	285 twin disc	305 twin disc	285 twin disc
Brake rear dia. (mm)	200	285 disc	200 **6**	285 disc	285 disc	285 disc
Front suspension	teles	teles	teles	teles	teles	teles
Spring movement (mm)	185	185 **7**	185 **8**	185	135	135
Rear type	s/a	s/a	s/a	s/a	Paralever	s/a
Spring movement (mm)	110	110	110	110	140	110
Petrol tank (litre)	21	21	21	22	22	22
Ignition system	electronic	electronic	electronic	electronic	electronic	electronic
Generator type	alternator	alternator	alternator	alternator	alternator	alternator
Output (watts)	460	460 **9**	460 **9**	460	460	460 **9**
Battery (volts)	12	12	12	12	12	12
Wheelbase (mm)	1516	1516	1516	1516 **10**	1565	1516
Seat height (mm)	810	810	810 **11**	810	780	810

Appendix

Model	K75C	K75S	K75	K100LT	K1	K75RT
Width (mm)	900	810	900	916	760	916
Length (mm)	2220	2220	2220	2220	2230	2220
Wet weight (lb/kg)	503/228	518/235	503/228	580/263 **12**	571/259	569/258
Power: bhp	75	75	75	90	100	75
@ rpm	8500	8500	8500	8000	8000	8500

1 1987—30 **2** 1991—2.81 with catalytic converter **3** 1991—Limited Edition 110/80V18 **4** 1989—130/90H18
5 1991—Limited Edition 140/80V17 **6** 1990—285 disc **7** 1987—135 **8** 1990—135 **9** 1994—700
10 1990—1511 **11** 1988-760 or 800 **12** 1991—624/283

Model	K100RS	K1100LT	K1100RS	K1200RS	F650	F650ST
Year from	**1990**	**1992**	**1993**	**1997**	**1994**	**1997**
Year to	**1992**	**1996**	**1996**	**1997**	**1997**	**1997**
Bore (mm)	67	70.5	70.5	70.5	100	100
Stroke (mm)	70	70	70	75	83	83
Capacity (cc)	987	1093	1093	1171	652	652
Inlet valve dia. (mm)	26.5	26.5	26.5	26.5	36	36
Exhaust valve dia (mm)	23	23	23	23	31	31
Compression ratio (to 1)	11.0	11.0	11.0	11.5	9.7	9.7
Gear ratio: top	1.61	1.61	1.61	1.51 (5th 1.70)	0.88	0.88
Gear ratio: 4th	1.88	1.88	1.88	1.96	1.05	1.05
Gear ratio: 3rd	2.30	2.30	2.30	2.39	1.31	1.31
Gear ratio: 2nd	2.96	2.96	2.96	3.02	1.75	1.75
Gear ratio: 1st	4.50	4.50	4.50	3.86	2.75	2.75
Axle ratio: solo	2.81 **1**	2.91 **2**	2.81	2.75	2.94	2.94
Front tyre (in.)	120/70V17	110/80VR18	120/70VR17	120/70ZR17	100/90S19	100/90-18
Rear tyre (in.)	160/60V18	140/80VR17	160/60VR18	170/60ZR17	130/80S17	130/80-17
Brake front dia (mm)	305 twin disc	305 twin disc	305 twin disc	305 twin disc	300 disc	300 disc
Brake rear dia. (mm)	285 disc	285 disc	285 disc	285 disc	240 disc	240 disc
Front suspension	teles	teles	teles	Telelever	teles	teles
Spring movement (mm)	135	135	135	115	170	170
Rear type	Paralever	Paralever	Paralever	Paralever	s/a	s/a
Spring movement (mm)	120	120	120	150	165	120
Petrol tank (litre)	22	22	22	21	17.5	17.5
Ignition system	electronic	electronic	electronic	electronic	electronic	electronic
Generator type	alternator	alternator	alternator	alternator	alternator	alternator
Output (watts)	460	460 **3**	700	720	280	280
Battery (volts)	12	12	12	12	12	12
Wheelbase (mm)	1564	1565	1565	1555	1480	1465
Seat height (mm)	800	810	800	770/800	810 **6**	785 or 735
Width (mm)	800	915	800	850	880	880
Length (mm)	2230	2250	2230	2250	2180	2160
Wet weight (lb/kg)	571/259	639/290	591/268	628/285	417/189 **7**	421/191
Power: bhp	100	100	100	130 **4**	48 **8**	48 **8**
@ rpm	8000	7500	7500	8750 **5**	6500 **9**	6500 **9**

1 1991—2.91 with catalytic converter **2** 1994—2.81 **3** 1994—700 **4** or 98 **5** or 7000 **6** 1997—800 or 750
7 1997—421/191 **8** or 34 **9** or 5700

Colours

For many years all standard BMW machines were finished in black and white lining to tank and mudguards, the only colour relief being the two tank badges. The black finish often extended to the wheel rims so that only the exhaustr system, handlebars and minor parts would be chrome-plated. After the war this continued up to 1969, when, with the introduction of the /5 series, came colour. Before that the only variations were specials created by owners and batches of police bikes which were sometimes finished in white with black lining. In the early postwar years variations of the normal black with white lining occurred as noted below:

1951
R51/3 and **R67**: wheel rim centres in silver, outer parts in black.

1952
R25/2 and **R68**, as **R67**.

1969
R50/5 and **R75/5**: silver grey tank and mudguards lined blue; frame, forks, headlamp shell, black.
R60/5: all black, white lining.

1972
Standard colours for tank and mudguard red with white lining, black with white lining or white with black lining. Options in metallic finish in Polaris (silver) or Curry (mustard) with black lining, or blue or green with white lining. Optional chrome-plated tank and battery side panels.

1973
Red or metallic green with white lining for tank and mudguards. No tank chrome panels.
/6 models: choice of black, white, red, green, blue, Curry or Polaris for tank, side panels and mudguards. Double line on tank and guards in white, except when finish in that colour, then in black. Frame, forks, headlamp in black.
/S: special hand finish in silver to deep smoked grey with gold lining for tank, side panels, guards and seat unit.

1974: As 1973.

1975
/S: TT silver smoke (as in 1973) or Daytona orange with red lining.
/6: Monza blue, Nürburg green, Bol d'Or red, Avüs black – all with white lining in 1973 style; also in Imola silver with black lining.

1976
/6 and **/S**: as 1975.
/7 and **S**: as **/6**.
RS: silver.

1977: As 1976.

1978
All **/7** models available in blue, orange or black.
R100S: metallic red.
R100RS: metallic silver or gold.
R45 and **R65**: metallic red with gold lining or metallic silver-beige with olive lining for tank, mudguards, seat tail and side panels. Black frame, headlamp shell and fork crowns.
R100SRS: white with tri-colour striping.

1979
R45 and **R65**: as 1978.
R80/7: black, blue, red or orange with contrast lining; black side panels.
R100T: as **R80/7** as option, standard in metallic red with metallic silver tank sides and side panels.
R100S: metallic red or dark red; black side panels.

R100RT: metallic brown and silver, the latter for the lower fairing and side panels.
R100RS: gold with black side panels, or in blue and silver. Silver for tank sides, lower fairing and side panels; blue for tank top, seat tail and upper fairing.

1980
R45 and **R65**: as 1978 plus additional colour in metallic bronco/anthracite with red lining.
R80/7: metallic bronco/anthracite, red or dark blue.
R100T: as **R80/7** and in red and silver as 1979.
R100S: as 1979.
R100RT: as 1979 or in metallic dark red.
R100RS: blue and silver as 1979, or in ,metallic silver-beige.
R80G/S: white tank, mudguards and side panels; black frame; matt black exhausts and bars; blue and dark blue patches on tank; orange dualseat.

1981
Tank, mudguards, side panels, seat pan and tail in colour, frame in black.
R45: blue, red or black.
R65: as **R45**, plus green.
R100: dark blue, silver or black.
R100CS: metallic black
R100RT: smoked red or smoked green.
R100RS: smoked red or smoked grey.
R80G/S: as 1980.
R65LS: red tank, mudguards, seat tail and fairing with white wheels or those items in silver with silver wheels. Black frame, seat, side panels. Matt black exhaust system, black bars. Fairing with black centre part from top of headlamp to instruments.

1982
R45, R65, R65LS, R100, R100T: as 1981.

R100CS: as 1981 plus smoked red.
R100RS: as 1981 plus smoked silver.
R80G/S: as 1980 plus Pacific blue in place of white, black seat,tank patches in red and light blue.

From the early 1980s BMW motorcycles were shown in the brochures in one or more colours which were usually red, silver, blue or black. Special paintwork was often available as an option and colours would vary between models from year to year. These variations could be enhanced by changes to the seat colour, sometimes with further options for some, or all, of a range.

There were special finishes used for the limited edition models and these would have extra pinstripes on wheels or fairings. The engines and power train could be part of a special finish with the most oustanding the bright yellow applied to the wheels and power train of the K1.

In all, a very far cry from the traditional black and white lining.

Model recognition points

1949
R24: rigid frame, new 250 engine, new gearbox.

1950
R25: plunger frame, R24 engine and gearbox.
R51/2: pre-war type enginewith two chain driven camshafts, alloy head, extended valve rocker supports, coil valve springs, two piece rocker covers held by strap, dynamo on top of crankcase, downdraught carburettors, gearbox improved as **R25**, teles, plunger,

1951
R25/2: as **R25**, minor changes.
R51/3: single gear driven camshaft,

pancake dynamo on crankshaft end, one-piece rocker covers, air filter contianer on gearbox shell top, plunger frame from **R51/2**.
R67: as **R51/3** with larger engine.

1952
R25/2: no change.
R51/3: twin leading shoe front brake.
R67: brake as **R51/3**, raised power output.
R68: new model as **R67**, only 2 ribs on rocker cover, larger air cleaner housing.

1953
No changes, **R25/2** discontinued late in year.

1954
R25/3: new model, air inlet under tank, no hand shift lever, 18in. wheels, full width hubs, damped forks.
Twins: two-way fork damping, fork gaiters, full width hubs.
R51/3 and **R68**: alloy rims; models discontinued late in year.

1955
R25/3: no changes, discontinued late in year.
R50: Earles forks, swinging fork, diaphragmclutch, three-shaft gearbox, twin leading shoe front brake in full width hub.
R69: as **R50** with 600 engine.
R26: Earles forks, swinging fork, single leading shoe front brake.
R67: continued for sidecar use, plunger frame.

1956
R26, **R50**, **R69**: no change.
R67: no change, discontinued late in year.
R60: as **R50** with **R67** engine.

1957 to **1959**
R26, **R50**, **R69**: no changes.

1960
R50: no change.
R26, **R69**: no change, discontinued late in year.
R60: became **R60/2** with more power and stronger crankshaft.
R27: new, rubber mounted engine, increased power, auto cam chain tension, **R26** cycle parts.
R50S: sports **R50**, timed breather, hydraulic steering damper.
R69S: sports **R60**, as **R50S**.

1961
R50: became **R50/2** with stronger crankshaft.
R27, **R50S**, **R60/2**, **R69S**: no change.

1962
All models: no change.

1963
R27, **R50/2**, **R60/2**: no change.
R50S, **R69S**: vibration damper added to crankshaft.
R50S: discontinued in year.

1964 to **1966**
R27, **R50/2**, **R60/2**, **R69S**: no changes.

1967
R27: no changes, discontinued in year.
R50/2: US version with telescopic forks, type **R50 US**.
R60/2: as **R50/2**, type **R60 US**, discontinued late in year.
R69S: as **R50/2**, type **R69 US**.

1968
R50/2, **R50 US**, **R69S**, **R69 US**: no changes.

1969
R50, **R69S** models: no changes, discontinued in year. New models introduced in year:
R50/5: 500 cc, **R60/5**: 600 cc, **R75/5**: 750 cc.
All alloy engine, camshafy below crankshaft, four speeds, new frame,

teles, s/a, drum brakes. Electric starter option for 500, standard on others.

1972

Smaller petrol tank, WM3 rear rim, revised dualseat with hand-rail, firmer rear dampening, fly-back propstand, lighrter flywheel, lowered axle ratio for **R75/5**.

1973

/5 models: longer swinging fork, increased wheelbase. Discontinued late in year.

/6 models: introduced September in 600, 750 and 900 cc forms plus sports 900. All with 5-speed gearbox, side panels, hydraulic steering damper, instrument panel with matching speedometer and rev-counter, warning lights between.

R90/S: sports 900, cockpit fairing, seat with base and tail, twin front discs.

R75/6 and **R90/6**: single front disc.

R60/6: drum front brake.

1974

All /6 models: capacity shown on side panels.

1975

All models: stronger forks, 17 mm front wheel spindle; drilled discs, 0 6 hp starter, new petrol taps, new electric switches, no kickstart lever as standard but available as option.

1976

Many internal changes, quietening rib on barrels, deeper oil sump, wider fork yokes, larger brake cylinders. Mid-year /6 range discontinued and replaced by /7 models.

All models: new rocker box covers, reinforced frame, shorter but thicker fins, 24-litre tank with recessed filler, no rear spring covers.

R60/7: disc front brake.

R75/7, R100/7: single disc.

R100S: double disc, cockpit fairing as

R90/S.

R100RS: full fairing, optional alloy wheels.

1977

R75/7: replaced in September by **R80/7**: new model in 55 or 50 bhp form, single disc front.

R100RS: September on — rear disc brake, alloy wheels standard.

1978

All models: remote gearchange, electronic rev-counter, handlebar cover, audible tiun signals, first aiid kit under saddle front.

R80/7 and **R100/7**: twin front discs.

R100S: alloy wheels standard and rear disc brake.

R60/7: discontinued late in year. New models from late in year:

R45 and **R65**: shorter stroke engines, same gearbox, torsional damper in shaft drive, alloy wheels, 18 in. front, disc front, drum rear, shorter fork and swing arm movement, shorter wheelbase, instrument console, new style switchgear, 22-litre tank.

SRS: S fairing, crash bars, mudflap, hazard lights; discontinued end year.

1979

R80, **R100** models: alloy wheels, **R65** type switches, drive shaft torsion damper.

R100/7 became **R100T** with extras as standard.

R100RT: new model with big touring fairing.

1980

No changes, **R80/7** discontinued late in year. **R80G/S**: added late in year; trail format, monolever rear suspension, disc front, 2 into 1 exhaust, small headlight, 21 in. front wheel, alloy rims, trail tyres.

R100T replaced by **R100**, and **R100S** by **R100CS** at end of year.

1981

All models except **R80G/S**: deeper sump, air filter housing as **R65**, electronic ignition, choke levers on bars, second exhaust balance pipe, new silencers, **R65** fork style common, fixed caliper for disc front brakes, master cylinder on bars.

R100RT: Nivomat standard.

R45 and **R65**: Longer wheelbase.

R65LS: introduced late in year, new style with cockpit fairing, styled seat and seat tail, new front mudguard, twin front discs, new type of cast alloy wheels, dropped bars, grab handles built into seat.

1982

As **1981**. During year **R80ST** with monolever rear suspension and **R80RT** with 800 cc engine in **R100RT** chassis added.

1983

As 1982. In September **K100**, **K100RS** and **K100RT** announced.

1984

R100 series dropped, late in year **R80ST** re-coded as **R80** and **R80RT** changed to monolever rear suspension.

1985

R80G/S Paris-Dakar model added, **R45** and **R65LS** dropped, **R65** changed to new version using **R80** chassis with 650 cc engine. Late in year **K75C** and **K75S** announced

1986

No changes

1987

R65, **R80**, **R80RT** and **R80G/S** continued.

R100RS relaunched with monolever rear suspension

K75, basic model, no fairing.

K75S Special, sports model plus

engine spoiler.

K100LT, de luxe tourer based on **K100RT** plus extra fitments.

K100RS special, sports suspension, special finish.

At end of year **R80G/S** and **K75S** Special dropped.

1988

R65, **R80**, **R80RT**, **R100RS**, **K75**, **K75C**, **K75S**, **K100**, **K100RS**, **K100RT** and **K100LT** continued.

R65GS, new model as old **R80G/S** with smaller, 27 hp, engine.

R80GS, Paralever rear suspension, new forks and wheels.

R100GS, new model as **R80GS**.

R100RT relaunched with monolever rear suspension.

K100RS special, sports suspension, special finish.

ABS option for **K100** models available.

At end of year **K75C** and **K100RT** dropped.

1989

R65 listed for Germany only with 27 hp engine.

R100GS Paris-Dakar, new model as standard **R100GS** plus extra to suit rally.

K1, new model with 4-valve engine, Paralever rear suspension, Marzocchi forks, Brembo brakes, ABS option, sports fairing, special paint finish.

Other models, as before.

1990

ABS option for **K75** models.

K100, limited market only.

K100RS, fitted with 4-valve K1 engine, Paralever rear suspension, Marzocchi forks, Brembo brakes.

K75RT, new model as old **K100RT** with 3-cylinder engine and **K100LT** fairing, built for USA and Spain only.

At end of year **R65GS** and **K100** dropped.

1991

R80GS, cockpit fairing as on Paris-Dakar model.

R100GS, as **R80GS**.

All twins, secondary air system (SAS) to reduce emissions.

K1 and **K100RS**, controlled catalytic converter option.

Other **K**-series, catalytic converter option.

K100LT Limited Edition, new model, better equipped than standard.

K75RT, on general release.

At end of year both versions of **K100LT** dropped.

1992

R80, **R80RT**, Marzocchi forks.

R100R, retro model, Paralever suspension, wire wheels, rounded valve covers.

K1100LT, larger-capacity replacement for **K100LT** with Paralever suspension and electric screen adjustment.

At end of year **R100RS** and **K100RS** dropped.

1993

R80R, as **R100R**.

R80 not listed for all countries

R80 models listed with 27 bhp engine in Germany.

R1100RS, new boxer engine, Telelever front suspension, Paralever rear, second-generation ABS, Brembo brakes.

K1100RS, larger-capacity replacement for **K100RS.**

K75RT, electric screen adjustment.

At end of year **R80** and K1 dropped.

1994

F650 Funduro, built by Aprilia, Rotax engine, chain drive model.

R1100GS, off-road model based on **R1100RS**.

R80 models, listed with 50 or 34 bhp engines.

R80, only listed for Germany and with 34 bhp engine.

R100R, twin front discs, SAS.

R100R Mystik, **R100R** in special finish.

All **K**-series, 700-watt alternator.

K1100 models, second-generation ABS.

K1100LT Special Edition, special finish and fittings.

At end of year **R80**, **R80RT**, **R80GS**, **R80R** and **R100GS** dropped.

1995

R1100R, roadster model using new boxer engine, Telelever, Paralever, alloy wheels, wire wheel option.

R850R, smaller version of **R1100R** with 70 or 34 bhp engine option.

Other models as before.

At end of year **R100R**, **R100GS-PD**, **R100RT**, **R100R** Mystik and **K75S** dropped.

1996

R850R, **R1100R**, **R1100GS**, **K75**, **K75RT** and **K1100RS**, no changes.

R1100RS, longer front wheel cover.

R1100RT, tourer with fairing, luggage system, **R1100RS** engine, Telelever, Paralever and ABS as standard.

K1100LT, ABS standard, windshield switch on left handlebar.

F650, centre stand fitted as stock.

During year **K75** and **K75RT** dropped, at end of year **K1100RS** and **K1100LT** Special Edition dropped.

1997

R850R, **R1100R**, **R1100GS** and **R1100RT**, no changes.

R1100RS, suspension strut changes.

F650, new fairing, screen, lower seat.

F650ST, new road model based on **F650**.

K1100LT, basic model not fitted with panniers or top box.

K1100LT Highline, replaced Special Edition.

K1200RS, new model in place of **K1100RS**, larger engine, six speeds, aluminium frame, Telelever, Paralever, alloy wheels.